Antoni Ligêza

Logical Foundations for Rule-Based Systems

Studies in Computational Intelligence, Volume 11

Editor-in-chief
Prof. Janusz Kacprzyk
Systems Research Institute
Polish Academy of Sciences
ul. Newelska 6
01-447 Warsaw
Poland
E-mail: kacprzyk@ibspan.waw.pl

Further volumes of this series
can be found on our homepage:
springer.com

Vol. 1. Tetsuya Hoya
Artificial Mind System – Kernel Memory Approach, 2005
ISBN 3-540-26072-2

Vol. 2. Saman K. Halgamuge, Lipo Wang (Eds.)
Computational Intelligence for Modelling and Prediction, 2005
ISBN 3-540-26071-4

Vol. 3. Bożena Kostek
Perception-Based Data Processing in Acoustics, 2005
ISBN 3-540-25729-2

Vol. 4. Saman K. Halgamuge, Lipo Wang (Eds.)
Classification and Clustering for Knowledge Discovery, 2005
ISBN 3-540-26073-0

Vol. 5. Da Ruan, Guoqing Chen, Etienne E. Kerre, Geert Wets (Eds.)
Intelligent Data Mining, 2005
ISBN 3-540-26256-3

Vol. 6. Tsau Young Lin, Setsuo Ohsuga, Churn-Jung Liau, Xiaohua Hu, Shusaku Tsumoto (Eds.)
Foundations of Data Mining and Knowledge Discovery, 2005
ISBN 3-540-26257-1

Vol. 7. Bruno Apolloni, Ashish Ghosh, Ferda Alpaslan, Lakhmi C. Jain, Srikanta Patnaik (Eds.)
Machine Learning and Robot Perception, 2005
ISBN 3-540-26549-X

Vol. 8. Srikanta Patnaik, Lakhmi C. Jain, Spyros G. Tzafestas, Germano Resconi, Amit Konar (Eds.)
Innovations in Robot Mobility and Control, 2005
ISBN 3-540-26892-8

Vol. 9. Tsau Young Lin, Setsuo Ohsuga, Churn-Jung Liau, Xiaohua Hu (Eds.)
Foundations and Novel Approaches in Data Mining, 2005
ISBN 3-540-28315-3

Vol. 10. Andrzej P. Wierzbicki, Yoshiteru Nakamori
Creative Space, 2005
ISBN 3-540-28458-3

Vol. 11. Antoni Ligêza
Logical Foundations for Rule-Based Systems, 2006
ISBN 3-540-29117-2

Antoni Ligêza

Logical Foundations for Rule-Based Systems

Second Edition

Springer

Professor Antoni Ligêza
Institute of Automatics
AGH - University of Science and Technology
Al. Mickiewicza 30
30-059 Cracow
Poland
e-mail: ligeza@agh.edu.pl

Originally published in Poland by AGH University of Science and Technology Press, Kraków, Poland

ISSN print edition: 1860-949X
ISSN electronic edition: 1860-9503

ISBN-13 978-3-642-06709-9 e-ISBN-13 978-3-540-32446-1

Springer is a part of Springer Science+Business Media
springer.com

Softcover reprint of the hardcover 2nd edition 2006

Preface

Thinking in terms of facts and rules is perhaps one of the most common ways of approaching problem definition and problem solving both in everyday life and under more formal circumstances. The best known set of rules, the *Ten Commandments* have been accompanying us since the times of Moses; the Decalogue proved to be simple but powerful, concise and universal. It is logically consistent and complete. There are also many other attempts to impose rule-based regulations in almost all areas of life, including professional work, education, medical services, taxes, etc. Some most typical examples may include various codes (e.g. legal or traffic code), regulations (especially military ones), and many systems of customary or informal rules.

The universal nature of rule-based formulation of behavior or inference principles follows from the concept of rules being a simple and intuitive yet powerful concept of very high expressive power. Moreover, rules as such encode in fact *functional* aspects of behavior and can be used for modeling numerous phenomena.

There are two main types of rules depending on their origin: there are objective, *physical rules* defined for us by *Nature* and there are subjective, *logical rules* defined by man. Physical rules describe certain natural phenomena and behavior of various systems; they are known by observation and experience, sometimes they can be proved, having objective nature they are independent of our will, they are universal and normally cannot be changed. Logical rules are those defined by man; they are usually subjective, local, subject to change if necessary. Physical rules describe possible behavior — they can be used in domains such as modeling, analysis and prediction of system behavior. Logical rules are usually aimed at shaping the behavior of man, society or machine. In any case definition of logical rules must respect the necessity of taking into consideration physical rules which cannot be violated — physical rules are superior with respect to logical ones.

Although rule-based systems are a tool omnipresent in science, technology and everyday life, their encoding, analysis and design are seldom a matter of deeper theoretical investigation; in most of the application areas they are just used (consciously or unconsciously) in a straightforward way, applied to solve specific problem without paying attention to issues such as their properties, language, optimization, etc.

The most thorough analyses of rules, inference, and rule-based systems were performed in the domain of logic[1]. Although rule-based inference is not the only possibility of reasoning, logical systems are mostly constructed as composed of axioms (facts) and inference rules. Theoretical properties of such systems, such as logical consistency and completeness are those recognized of primary importance and investigated.

The rule-based approach for knowledge representation and reasoning has been adapted from logic to *Artificial Intelligence* (AI) and *Knowledge Engineering* (KE) [39, 44, 125]. The so-called *production systems* [125] or rule-based systems [44, 46] are sets of rules imitating logical implication. Even after years of investigation of various other formalisms, rules proved to be generic, core and very universal knowledge representation tool for the widest possible spectrum of applications.

Rule-Based Systems (RBS) constitute a powerful tool for specification of knowledge in design and implementation of knowledge-based systems (KBS) in applied Artificial Intelligence and Knowledge Engineering. They provide also a universal programming paradigm for domains such as system monitoring, intelligent control, decision support, situation classification, system diagnosis and operational knowledge encoding. Apart from off-line expert systems and deductive data-bases, one of the most useful and successful applications consists in development of wide spectrum of control and decision support systems [48].

In its basic version (considered here) a RBS for control or decision support consists of a single-layer set of rules and a simple inference engine; it works by selecting and executing a single rule at a time, provided that the preconditions of the rule are satisfied in the current state. Possible applications include direct control and monitoring of dynamical processes [66], meta-level control (the so-called expert control), implementation of the low level part of any-time reactive systems, generation of operational decision support, etc. A RBS named *Kheops* [42], being one classical example of such systems was applied in the TIGER system [88, 126] developed for gas turbine monitoring. Many successful applications are reported in [51] and in [48].

The expressive power and scope of potential applications combined with modularity make RBS a very general and readily applicable mechanism. However, despite a vast spread-out in working systems, their theoretical analysis seems to constitute still an open issue with respect to analysis, design methodologies and verification of theoretical properties. Assuring *reliability*, *safety*,

---

[1] The book focuses on classical First-Order Predicate Calculus, Resolution, theorem proving and following tools, such as PROLOG programing language and forward chaining rule-based systems. The issues of λ-calculus and LISP are not mentioned in this book.

quality and *efficiency* of rule-based systems requires both theoretical insight and development of practical tools. The general qualitative properties are translated into a number of more detailed characteristics defined in terms of logical conditions.

In fact, in order to assure safe and reliable performance, such systems should satisfy certain formal requirements, including completeness and consistency. To achieve a reasonable level of efficiency (quality of the knowledge-base) the set of rules must be designed in an appropriate way. Several theoretical properties of rule-based systems seem to be worth investigating, both to provide a deeper theoretical insight into the understanding of their capacities and assure their satisfactory performance, e.g. *reliability* and *quality* [3, 48, 101, 103, 107, 123]. Some most typical issues of theoretical verification include satisfaction of properties such as *consistency*, *completeness*, *determinism*, *redundancy*, *subsumption*, etc. (see [3, 81, 101]). Several papers investigate these problems presenting particular approaches [25, 103, 107, 123]. A selection of tools is presented in [109]. Some modern approaches include [6, 49, 132]. An interesting extension concerns analysis and verification of time-dependent systems, especially real-time systems [17].

RBS provide a powerful tool for knowledge specification and development of practical applications. However, although the technology of RBS becomes more and more widely applied in practice, due to its relationship to first-order logic and sometimes complex rule patterns and inference mechanisms, they are still not well-accepted by industrial engineers. Further, the 'correct' use of them requires much intuition and domain experience, and knowledge acquisition still constitutes a bottleneck for many potential applications. Software systems for development of RBS are seldom equipped with tools supporting design of the knowledge-base; for some exceptions see [1, 4]. A recent, new solution is proposed in [141]. However, a serious problem follows from the fact that a complete analysis of properties remains still a problem, especially one supporting the design stage rather than the final verification. This is particularly visible in case of more powerful knowledge representation languages, such as ones incorporating the full first order logic formalism.

Contrary to RBS, *Relational Data Base Systems* (RDBS) [23, 30, 38, 131] offer relatively simple, but matured data manipulation technology, employing widely accepted, intuitive knowledge representation in tabular form. It seems advantageous to make use of elements of this technology for simplifying certain operations concerning RBS. Note that from practical point of view any row of a RDBS table can be considered as a rule, provided that at least one attribute has been selected as an output (and there is a so-called *functional dependency* allowing for determination of the value of this attribute on the base of some other attributes). Thus, it seems that merging elements of RBS and RDBS technologies can constitute an interesting research area of potential practical importance.

There exist numerous books and papers presenting the rule-based systems as a methodology for knowledge representation and inference with applications. Some best examples of such books include classical positions, such as [39, 43] and [125] with respect to logical foundations [44, 46] and [117] covering classical presentation, and [130] and [48] with respect to applications. A comprehensive, multi-author presentation of the most wide spectrum of issues concerning rule-based systems is perhaps covered by the handbook edited by J. Liebowitz [51]. Yet another, interesting and new one is the work [102]. With respect to real-time systems the specific issues are presented in [17]. All these positions cover certain aspects of rule-based systems and present interesting and useful material on that methodology. However, one main drawback common to such positions is that trying to be attractive they present the material at rather popular level without going to more difficult details. They also omit many particular issues important in practical implementations and applications. For example, no textbook on rule-based systems explore the relationship between RDBS and attributive rule-based systems. No books point to similarities in both of the technologies and analyze possibilities of at least partial merging of them. Last but not least, they are full of repetitions of a basic, well-known material which is presented in similar way in other textbooks. Analyses and discussions focused on selected theoretical or application-oriented details, providing in-depth analysis of more specific problems can hardly be met in the books addressed to a wider audience.

This book addresses the methodology of rule-based systems in a relatively complete and perhaps a bit complex way. The main aim is to present the rule-based systems from logical perspective as viewed by the Author. Certain Author's concepts concerning rule-based systems are described in details. Although the primary concern of this book may seem to be well-explored in the domain literature, both the structure and the contents of the book attempt at keeping individual, Author-shaped character, and present personal experience of both theoretical and practical nature.

The concept of the book is as follows: to present in a single volume a spectrum of knowledge concerning rule-based systems, as understood in knowledge engineering, but with going into details uncovered by other books on that topic. The book covers areas such as: logical foundations of rule-based systems (including knowledge representation and inference with propositional, attribute-based and first-order logic), knowledge representation, inference and inference control in rule-based systems (including extended forms of rules and specialized inference control mechanisms), definitions and verification of formal properties of rule-based systems assuring the correct work of them and finally design issues (covering systematic design approach combined with on-line verification). The discussion is presented at the conceptual level, then logical definitions are systematically introduced and practical implementation-oriented solutions are provided. In several, most distinctive cases, the discussion is continued into details of implementation illustrated by working solutions in PROLOG.

Contrary to majority of textbooks on Artificial Intelligence, Knowledge Engineering and Rule-Based Systems, which attempt at concise and comprehensive presentation of a mixture of approaches sometimes completely different from one another, this books follows in a consequent way a single line of presentation: it starts the lecture at the very beginning — the propositional calculus. It goes through logical languages for knowledge representation, inference rules, principles and details of rule-based systems, until design and verification issues. It offers also practical solutions illustrated with PROLOG code excerpts. Hence, apart from introducing and explaining many technical issues it provides practical instructions how to implement the ideas in an efficient way.

The book presents also some *ideology* concerning design and development of rule-based systems for practical applications. The principal lines distinguishing the presented material can be summarized as follows:

- *knowledge algebraization* — although rule-based systems were born in the area of logic and inherit often the logical terminology, notation, and inference mechanisms, for practical applications they can be made 'as algebraic as possible', close to well-known and very efficient Relational Database technology; this means that rules represented in attributive languages can be presented in tabular form easy to analyze and manipulate by algebraic means;
- *hierarchical organization of knowledge* — the initial problem-space can be divided into local, specific *contexts*, each of them having precise logical definition, and the contexts are organized in a tree-like structure; the design of the system and the final system components can reflect the problem structure what makes it easier to analyze and design the rule-based system thanks to decomposition into smaller parts;
- *formalization of design and verification* — whenever possible, the design and verification process should be formal and the designed system should provide required functionality preserving important characteristics, such as consistency, completeness, etc.; in order to assure those characteristics an attempt to put forward algebraic and graphical knowledge representation enabling easy design (which should be 'almost mechanical') is undertaken.

With respect to the principal guidelines assumed and presented above, a number of specific solutions were proposed. The most important, original issues addressed in this book include the following:

- presentation of logical languages for encoding rule-based systems with special attention paid to attribute-based languages; four types of such languages were introduced and specific inference mechanisms were presented;
- presentation of logical inference method called *backward dual resolution* (or *dual resolution* for short) which is especially convenient for analysis of completeness and reduction of rules; it can also be applied in first-order logic based systems for proving satisfaction of rules preconditions in case of complex DNF-like formulae;

- proposal of extended, frame-like form of inference rules containing numerous components and allowing for dynamic memory modification and encoding inference control in declarative rules;
- knowledge representation method in the form of *Extended Tabular Systems* (XTT) where knowledge is encoded in tabular components linked into a tree structure for efficient control, and where non-atomic values of attributes are allowed;
- logical definitions and practical approach to verification of certain important formal properties of rule-based systems;
- a proposal of new rule-based systems designing paradigm incorporating graphical knowledge representation and on-line verification;
- last but not least, practical aspects of encoding the ideas in PROLOG as a meta-level code.

The tabular systems discussed in this book can also be used as extended RDB paradigm for unconditional knowledge specification. In such a way instead of extensional data specification with atomic values of attributes, their intensional definition can be provided. In the basic case, set and interval values of attributes can be used to cover a number of specific cases. Depending on the knowledge representation language, also more complex structures (e.g. records, objects, terms) can be used. In such a way *data patterns*, *data covers* or *data templates* can be defined. Both representation and analysis can be then much more concise and efficient.

The organization of this book is as follows. The book is divided into five parts, each of them further divided into several chapters. The main parts present material on (i) logical foundations of rule-based systems (Part I), (ii) principles of rule-based systems structures, knowledge representation languages, inference and inference control (Part II), (iii) verification of formal properties of rule-based systems (Part III), and (iv) design methodology for efficient development of such systems and an extended example (Part IV). Part V presents concluding remarks and information on selected systems and web resources.

This book is addressed to researchers, students and engineers interested in the rule-based systems technology in all aspects, including theoretical foundations, languages, knowledge representation, inference, design and verification. It should serve as a material for self-study, both systematic one, from the bases, and as well as auxiliary material proposing more detailed, specific concepts and solutions provided on demand. I hope it will provide an interesting and useful material and perhaps a source of inspiration to those involved in knowledge engineering theory and practice.

Acknowledgment

The ideas and solutions presented in the book are the result of a long-term research work and teaching in the domain of Knowledge Engineering, including participation in several projects, many international conferences and workshops and work as a visiting professor at foreign universities and research institutions. During that time the Author met many people who supported his work in one or another way.

I would like to address my particular thanks to my teachers and professors who have had significant influence on shaping my research interests as well as attitude to research and introduced me to *System Theory* and systematic approach to analysis of research problems.

Special thanks are due to Prof. Henryk Górecki, Ph.D., the creator, teacher and Head of regular Ph.D. studies at the Faculty of Electrical Engineering, Automatics, Computer Science and Electronics at the AGH-UST[1] in Cracow; as a participant during the years 1980–1983 I was able to start my research and conclude the study with obtaining the Ph.D. in 1983. Prof. Henryk Górecki was the creator and for many years the Head of the Institute of Automatics at the AGH-UST. I would also like to thank him for scientific and friendly atmosphere at work and enabling me to continue my career in the domain of Artificial Intelligence, which at that time did not seem to be a very promising area.

Specially cordial thanks are due to Prof. Zbigniew Zwinogrodzki, Ph.D. who introduced me into the meanders of the fascinating world of logic and was taking care of my early research into logical knowledge representation and planning systems as well as for his further work towards preparing the bases for my Ph.D. Thesis; here I would like to express my gratitude for peer reviews of my early awkward writing attempts and for constructive criticism which helped in improving this book.

[1] AGH University of Science and Technology, located in Cracow (Kraków), Poland is a leading and one of the biggest technical universities in Poland.

I would also like to thank Prof. Ryszard Tadeusiewicz, Ph.D., who introduced me to 'other-than-logical' Artificial Intelligence methods and supervised the final stages of my Ph.D. thesis.

Many thanks are due to Prof. Tomasz Szmuc, Head of the Computer Science Laboratory at AGH-UST, where I was able to continue my research until today thanks to favorable atmosphere and high research standards which encouraged friendly competition and helped keeping on and concentrate on research even in these difficult times.

My research was influenced also during long and short-term periods of work abroad, especially in the LAAS-CNRS in Toulouse (1992, 1996), at the University of Nancy I (1994), at the University of Palma de Mallorca (1994, 1995), at the University of Girona (1997, 1998), and at the University of Caen (2004). I would like to thank Prof. Josep Aguilar Martin, Prof. Jean Paul Haton, Prof. Josep Lluis de la Rosa y Esteva, Dr Louise Travé-Massuyès, Dr Maroua Bouzid and Dr Pilar Fuster Parra for their kind help both with respect to research support and cooperation, as well as with the organizing of visits.

I am indebted to Dr Jacek Martinek from Technical University of Poznań for his friendly criticism of some of my research papers and the further work concerning habilitation (which turned out to be constructive, I hope); it helped me to improve my papers and eliminate many errors.

Many thanks to my Colleagues with whom we spent several years on research and preparing papers and Ph.D. theses: Dr Grzegorz Jacek Nalepa, for his inspiration and implementation of practical prototype system MIRELLA, as well as for his inestimable, continuous help with configuration, installation and repair of Linux software and LaTeX; I also would like to thank him for his courtesy concerning permission to incorporate in this book some elements from his Ph.D. Thesis, including presentation of design approaches (especially the XTT-based one) and the technical material enclosed in the appendices.

Many thanks to Dr Igor Wojnicki for designing and implementing the OSIRIS system [141] and Dr Marcin Szpyrka for multidimensional support.

Many thanks to Marek Kapłański, M.Sc. for his continuous support with maitaining my PC-s and the software running on it.

There is also a place to acknowledge the support of the KBN (the Polish State Committee on Scientific Research and currently the Polish Ministry of Science and Informatization). The research presented in this book was carried out with financial support of the KBN Grant No.: 8 T11C 019 17 during the years 1999–2002 — the ***Regulus*** project[2], and currently under the Grant No.: 4 T11C 035 24 — the ***Adder*** project[3].

[2] For more information on this project entitled *Formal Methods and Tools for Computer-Aided Analysis and Design of Databases and Knowledge-Based Systems* see the WWW page `http://regulus.ia.agh.edu.pl`.

[3] For more information on this project entitled *Application of Formal Methods to Support Development of Software for Real-Time Systems* see the WWW page `http://adder.ia.agh.edu.pl`.

Many thanks to Anna and Patrick Sarker who devoted their time to help me to improve the style and to correct my linguistic errors, inevitable for someone who is not a native speaker.

Last but not least, I am indebted to my wife, Ewa, for her help, patience, and continuous support during the years spent on research and the writing of this book. Moreover, she has always been a source of inspiration for me, specially with her unique way of reasoning, which does not lend itself for presentation in any formal, logical way, but still comes up with perfect solutions for problems in real time. Many thanks to my beloved daughters, Marianna and Magdalena for enabling and encouraging this long-term work. Many most warm thanks to my Parents.

Contents

Part I Logical Foundations of Rule-Based Systems

Part II Principles of Rule-Based Systems

Part V Closing Remarks and Appendices

Part I

Logical Foundations of Rule-Based Systems

1

Propositional Logic

Propositional Calculus is the simplest logical system, both with respect to syntax as well as semantics. It uses simple logical formulae constructed from propositional symbols and logical connectives only; no individual variables nor quantifiers are allowed. Simultaneously, it introduces many basic ideas incorporated in any more advanced logical systems. It can also serve as a basic model for rule-based systems.

The name *Propositional Calculus* (or *Propositional Logic*) comes from the fact that this kind of logic is limited to use of *propositions* as the only means for expressing knowledge about facts in some world under consideration. A *statement* or a *proposition* is any finite declarative sentence. In classical logic any proposition is either *true* or *false*, although in particular situation its current logical value may be unknown.

1.1 Alphabet of Propositional Calculus

The alphabet of any formal language consists of a set of items (letters, symbols) which are legal in this language. The alphabet of Propositional Calculus consists of symbols denoting propositions and logical connectives (logical functions). As auxiliary symbols parentheses are also allowed. Moreover, two special symbols for denoting a formula which is always true, say $\top$, and a formula which is always false, say $\bot$ will be necessary. The complete alphabet is specified as follows.

Definition 1. *The alphabet of Propositional Calculus consists of:*

- *a set of propositional symbols*

$$P = \{p, q, r, \ldots, p_1, q_1, r_1, \ldots, p_2, q_2, r_2, \ldots\},$$

- *a set of logical connectives, i.e.* $\wedge$ *(conjunction),* $\vee$ *(disjunction),* $\neg$ *(negation),* $\Rightarrow$ *(implication) and* $\Leftrightarrow$ *(equivalence),*

A. Ligęza: *Logical Foundations for Rule-Based Systems*, Studies in Computational Intelligence (SCI) **11**, 3–35 (2006)
www.springerlink.com

- *two special symbols, i.e.* $\top$ *denoting a formula always true, and* $\bot$ *denoting a formula always false.*

Moreover, parentheses are used if necessary.

For practical applications propositional symbols can be assigned some specific meaning; depending on the context and current needs any such symbol can be used to denote some declarative sentence having precisely defined meaning. The sentence can be evaluated to be true or false in the world under consideration. In this way the symbol is assigned the *truth-value*. However, when analyzing certain set of formulae one may think in abstract terms and use propositional symbols without any specific meaning assigned. In such a case it is said that such symbols are *propositional variables* — their meaning is unknown and the only restriction is that after assigning a specific interpretation they can be evaluated as true or false ones.

In order to assign some precise meaning to propositional symbol p, for example 'It is cold', the following notation can be used

$$p \stackrel{\text{def}}{=} \text{'It is cold'} .$$

Note that any propositional variable can be assigned a unique meaning only. Further, as we shall see, two propositional variables may be *independent* or *dependent* on each other. For intuition, they are independent if the assigned interpretations are independent; in such a case the variables can take logical values independently on each other. They are dependent if the interpretation of one of them known to be true (false) implies that the other interpretation is known.

1.2 Syntax of Propositional Logic

The only legal expressions of Propositional Logic are *well-formed formulae* (or formulae, for short), i.e. specific expressions constructed from the symbols of the alphabet according to certain rules. A formula is an expression that can be assigned a logical value (true or false). The formal definition of propositional logic formulae is specified by defining the set of formulae *FOR* in the following way.

Definition 2 (Propositional Logic formulae). *Let* P *denote the set of propositional symbols. The set of all propositional logic formulae* FOR *is defined inductively as follows:*

- *two special formulae* $\top \in FOR$ *and* $\bot \in FOR$*;*
- *for any* $p \in P$, $p \in FOR$*;*
- *if* $\phi \in FOR$ *then* $(\neg\phi) \in FOR$*;*
- *if* $\phi, \psi \in FOR$ *then* $(\psi \wedge \phi) \in FOR$, $(\psi \vee \phi) \in FOR$, $(\psi \Rightarrow \phi) \in FOR$ *and* $(\psi \Leftrightarrow \phi) \in FOR$*;*
- *no other item belongs to* FOR*.*

The elements of $P \cup \{\top, \bot\}$ are also called *atomic formulae* or *atoms* for short. All the formulae are constructed from atoms connected with use of logical connectives. Despite the use of parentheses, the following order (priority) of logical connectives is assumed:

- negation ($\neg$),
- conjunction ($\wedge$),
- disjunction ($\vee$),
- implication ($\Rightarrow$),
- equivalence ($\Leftrightarrow$).

Thus in certain cases parentheses can be omitted. For example, $(\neg\phi)\wedge(\neg\psi)$ can be simplified to $\neg\phi\wedge\neg\psi$; similarly, $(\neg\phi)\vee(\neg\psi)$ can be simplified to $\neg\phi\vee\neg\psi$. Further, $\phi \vee (\psi \wedge \varphi)$ can be simplified to $\phi \vee \psi \wedge \varphi$. However, $\phi \wedge (\psi \vee \varphi)$ is different from $\phi \wedge \psi \vee \varphi$.

In case of more complex formulae it may be useful to put parentheses in order to show the real structure of the formula.

Note that the above definition is recursive in fact. Having a well-formed formula one can replace any propositional symbol (or a formula symbol) with another well-formed formula; in this way a new well-formed formula is obtained. Such a replacement will be called *substitution*.

Let ϕ and φ be two formulae. The replacement of ϕ by φ is denoted as ϕ/φ. The simultaneous replacement of $\phi_1, \phi_2, \ldots, \phi_n$ with $\varphi_1, \varphi_2, \ldots, \varphi_n$ is denoted as $\{\phi_1/\varphi_1, \phi_2/\varphi_2, \ldots, \phi_n/\varphi_n\}$.

1.3 Semantics of Propositional Logic

In order to evaluate any Propositional Logic formula it is necessary to assign a meaning to its symbols. In this way the *interpretation* of propositional variables is specified. Having defined the interpretation it is possible to decide whether the statements are true or false. This process of establishing relationship among symbols and their meaning is named assigning an interpretation to propositional formulae. Assigning the truth value to a formula consists of evaluating the truth value of its components and the whole formula at the end.

From mathematical point of view, in order to assign truth value to propositional symbols one has to define an appropriate mapping I. Let P be the set of propositional symbols, and let $\{\mathbf{T}, \mathbf{F}\}$ denote the set of truth values (true and false, respectively).

Definition 3. *An interpretation I is any function of the form*

$$I \colon P \longrightarrow \{\mathbf{T}, \mathbf{F}\} \,. \tag{1.1}$$

In case $I(p) = \mathbf{T}$ we shall say that p is true under interpretation I. This can be also written as

$$\models_I p \,,$$

which is read as 'p is satisfied under interpretation I'. On the other hand, if $I(p) = \mathbf{F}$ we shall say that p is false under interpretation I. This can be also written as

$$\not\models_I p\,,$$

which is read as 'p is false (unsatisfied) under interpretation I.

The definition of interpretation is extended over the set of all formulae FOR in the following way.

Definition 4. *Let I be an interpretation of propositional symbols in P. Let FOR be the set of all formulae defined with symbols of P, and let ϕ, ψ and φ be any formulae, $\phi, \psi, \varphi \in FOR$. The truth value of formulae in FOR is defined as follows:*

- $I(\top) = \mathbf{T}$ ($\models_I \top$),
- $I(\bot) = \mathbf{F}$ ($\not\models_I \bot$),
- $\models_I \neg\phi$ *iff* $\not\models_I \phi$,
- $\models_I \psi \wedge \varphi$ *iff* $\models_I \psi$ *and* $\models_I \varphi$,
- $\models_I \psi \vee \varphi$ *iff* $\models_I \psi$ *or* $\models_I \varphi$,
- $\models_I \psi \Rightarrow \varphi$ *iff* $\models_I \varphi$ *or* $\not\models_I \psi$,
- $\models_I \psi \Leftrightarrow \varphi$ *iff* $\models_I (\psi \Rightarrow \varphi)$ *and* $\models_I (\varphi \Rightarrow \psi)$.

According to the above rules any well-formed formula of FOR can be assigned its truth value in a unique way, provided that initial interpretation of propositional symbols is known. For practical purposes, the rules of the above definition are usually presented in a readable tabular form.

The table defining negation is as follows

ϕ	$\neg\phi$
F	**T**
T	**F**

The table defining conjunction is as follows

ϕ	φ	$\phi \wedge \varphi$
F	**F**	**F**
F	**T**	**F**
T	**F**	**F**
T	**T**	**T**

The table defining disjunction is as follows

ϕ	φ	$\phi \vee \varphi$
F	**F**	**F**
F	**T**	**T**
T	**F**	**T**
T	**T**	**T**

The table defining implication is as follows

ϕ	φ	$\phi \Rightarrow \varphi$
F	**F**	**T**
F	**T**	**T**
T	**F**	**F**
T	**T**	**T**

Finally, the table defining equivalence is as follows

ϕ	φ	$\phi \Leftrightarrow \varphi$
F	**F**	**T**
F	**T**	**F**
T	**F**	**F**
T	**T**	**T**

Below some basic theoretical properties of well-formed formulae are given.

Definition 5. *A formula $\phi \in FOR$ is consistent (satisfiable) iff there exists an interpretation I under which the formula is satisfied, i.e.*

$$\models_I \phi\,.$$

Definition 6. *A formula $\phi \in FOR$ is falsifiable (invalid) iff there exists an interpretation I under which the formula is false, i.e.*

$$\not\models_I \phi\,.$$

Most of the formulae are both consistent and falsifiable; their truth value depends on the specific interpretation. However, there are some formulae the value of which is the same disregarding the interpretation.

Definition 7. *A formula $\phi \in FOR$ is inconsistent (unsatisfiable) iff there does not exist any interpretation I under which the formula is satisfied, i.e.*

$$\not\models_I \phi$$

for any possible interpretation I. This will be denoted shortly as $\not\models \phi$.

A typical example of an unsatisfiable formula is one of the form $\phi = \varphi \wedge \neg\varphi$. No matter how complicated φ is, ϕ is always false.

Definition 8. *A formula $\phi \in FOR$ is valid (is a tautology) iff for any interpretation I the formula is satisfied; this is formally written as*

$$\models \phi\,.$$

A typical example of a tautology is a formula of the type $\phi = \varphi \vee \neg\varphi$. No matter how complicated φ is, ϕ is always true.

The following lemma specifies some obvious observations following from the definitions [16].

Lemma 1. *The well-formed formulae satisfy the following properties:*

1. *A formula is tautology iff its negation is unsatisfiable; a formula is unsatisfiable iff its negation is tautology.*
2. *If a satisfiable formula is true under a certain interpretation, then its negation is false under the same interpretation; if a falsifiable formula is false under a certain interpretation, then its negation is true under the same interpretation.*
3. *Any tautology is consistent; any inconsistent formula is falsifiable.*

Two logical formulae can be compared with respect to the interpretations under which they are satisfied. Roughly speaking one of them may be *more general* than the other which is *more specific.* A more general formula is satisfied under any interpretation satisfying the more specific formula. A more general formula is also said to *logically follow* from the less general one, while the more specific formula *logically entails* the more general one.

Definition 9 (Logical consequence). *Let $\phi, \varphi \in FOR$ are any formulae. Formula φ logically follows from formula ϕ iff for any interpretation I satisfying ϕ, I also satisfies φ. This will be written shortly as*

$$\phi \models \varphi\,. \tag{1.2}$$

If (1.2) holds, we shall also say that φ is a *logical consequence* of ϕ.

Two logical formulae can be different but simultaneously taking the same logical value under any interpretation — such formulae will be said to be *logically equivalent.*

Definition 10 (Logical equivalence). *Formulae $\phi, \varphi \in FOR$ are logically equivalent iff for any interpretation I there is:*

$$\models_I \phi \qquad \textit{iff} \qquad \models_I \varphi\,. \tag{1.3}$$

In this case we shall write $\phi \models \varphi$ and $\varphi \models \phi$, or shortly, $\phi \equiv \varphi$.

The semantics of certain formulae can be defined with use of a certain basic set of logical connectives, e.g. negation and disjunction or negation and conjunction. For convenience, usually negation, conjunction and disjunction are used.

Below, some most common examples are presented:

- $\phi \Rightarrow \psi \equiv \neg\phi \vee \psi$,
- $\phi \Leftrightarrow \psi \equiv (\phi \Rightarrow \psi) \wedge (\phi \Leftarrow \psi)$,
- $\phi | \psi \equiv \neg(\phi \wedge \psi)$ — the so-called Sheffer function or `NAND`; another notation is $\overline{\phi \wedge \psi}$,
- $\phi \downarrow \psi \equiv \neg(\phi \vee \psi)$ — the so-called Pierce function or `NOR`; another notation is $\overline{\phi \vee \psi}$,
- $\phi \oplus \psi \equiv (\neg\phi \wedge \psi) \vee (\phi \wedge \neg\psi)$ — exclusive-OR function or `EX-OR`,
- $\neg\phi \vee \psi$ and $\phi \vee \neg\psi$ — asymmetric difference functions.

Generally, for n input propositional variables as many as 2^{2^n} different functions specifying logical connectives can be defined; so, for $n = 2$ there exist 16 different possibilities.

1.4 Rules for Transforming Propositional Formulae

Propositional formulae can be transformed from their initial form to another one which is logically equivalent. It is important to specify the legal transformations, i.e. the ones preserving logical equivalence. Transformation to other equivalent form is important for analysis and comparison of formulae.

The typical set of transformation rules is given below:

- $\neg\neg\phi \equiv \phi$ — double negation rule,
- $\phi \wedge \psi \equiv \psi \wedge \phi$ — commutativity of conjunction,
- $\phi \vee \psi \equiv \psi \vee \phi$ — commutativity of disjunction,
- $(\phi \wedge \varphi) \wedge \psi \equiv \phi \wedge (\varphi \wedge \psi)$ — associativity of conjunction,
- $(\phi \vee \varphi) \vee \psi \equiv \phi \vee (\varphi \vee \psi)$ — associativity of disjunction,
- $(\phi \vee \varphi) \wedge \psi \equiv (\phi \wedge \psi) \vee (\varphi \wedge \psi)$ — distributivity of conjunction with regard to disjunction,
- $(\phi \wedge \varphi) \vee \psi \equiv (\phi \vee \psi) \wedge (\varphi \vee \psi)$ — distributivity of disjunction with regard to conjunction,
- $\phi \wedge \phi \equiv \phi$ — idempotency of conjunction,
- $\phi \vee \phi \equiv \phi$ — idempotency of disjunction,
- $\phi \wedge \bot \equiv \bot$, $\phi \wedge \top \equiv \phi$ — identity laws for conjunction,
- $\phi \vee \bot \equiv \phi$, $\phi \vee \top \equiv \top$ — identity laws for disjunction,
- $\phi \vee \neg\phi \equiv \top$ — excluded middle law,
- $\phi \wedge \neg\phi \equiv \bot$ — inconsistency law,
- $\neg(\phi \wedge \psi) \equiv \neg(\phi) \vee \neg(\psi)$ — De Morgan's law,
- $\neg(\phi \vee \psi) \equiv \neg(\phi) \wedge \neg(\psi)$ — De Morgan's law,
- $\phi \Rightarrow \psi \equiv \neg\psi \Rightarrow \neg\phi$ — contraposition law,
- $\phi \Rightarrow \psi \equiv \neg\phi \vee \psi$ — definition of implication with disjunction and negation (elimination of implication).

1.5 Applications

The definition of interpretation as well as the transformation rules can be applied to verify some properties of formulae. There are the following main issues which can be checked either with use of examining possible interpretations or with use of successive formula transformation:

- tautology verification — checking if a formula is tautology,
- unsatisfiability verification — checking if a formula is unsatisfiable,
- logical equivalence verification — checking if two formulae are logically equivalent,

- logical consequence verification — checking if a formula logically follows from another formula,
- satisfiability verification — checking if a formula is satisfiable.

Basically, there are two different approaches to the problems presented above. The first approach is based on examining all possible interpretations; in certain situations some interpretations can be omitted. In the other approach the check is performed by applying the transformation rules. Below a simple example of the two approaches is shown.

Example. Consider the following formula

$$\phi = ((p \Rightarrow r) \wedge (q \Rightarrow r)) \Leftrightarrow ((p \vee q) \Rightarrow r)\ .$$

The problem is to check if the formula is tautology. Let us apply the first approach based on checking of all (2^3) possible interpretations.

For simplicity we change slightly the notation: instead of **T** we shall write 1 and instead of **F** we shall write 0. The process of checking all interpretations can be presented in a transparent way in the following tabular form, the so-called *logical matrix* of the formula (Table 1.1).

Table 1.1. Logical matrix for formula $\phi = ((p \Rightarrow r) \wedge (q \Rightarrow r)) \Leftrightarrow ((p \vee q) \Rightarrow r)$

p	q	r	$p \Rightarrow r$	$q \Rightarrow r$	$(p \Rightarrow r) \wedge (q \Rightarrow r)$	$(p \vee q) \Rightarrow r$	ϕ
0	0	0	1	1	1	1	1
0	0	1	1	1	1	1	1
0	1	0	1	0	0	0	1
0	1	1	1	1	1	1	1
1	0	0	0	1	0	0	1
1	0	1	1	1	1	1	1
1	1	0	0	0	0	0	1
1	1	1	1	1	1	1	1

The three leftmost columns specify all the eight possible different interpretations (since there are three input propositional variables, there are as many as $2^3 = 8$ different interpretations). For any interpretation we have to evaluate the components of the formula using the rules of Definition 4 and the following tables. Finally we evaluate ϕ (the last column). Since disregarding the interpretation the logical value of ϕ always equals true, the formula is tautology.

Note that in an analogous way logical equivalence of formulae can be verified; in fact, formulae specifying columns 6 and 7 are checked to be logically equivalent (for distinguishing them from the other previous columns, they are put in boldface). The same applies to verifying logical consequence, unsatisfiability, etc.

Now let us apply the second method based on logical transformation rules. For simplicity we shall keep the $\Leftrightarrow$ connective and transform the components on the left and on the right hand side. By elimination of implication the initial formula can be transformed into the following form

$$\phi \equiv ((\neg p \vee r) \wedge (\neg q \vee r)) \Leftrightarrow (\neg(p \vee q) \vee r) .$$

enddocumentNow, after applying the rule of distributivity to the left hand side we have

$$\phi \equiv ((\neg p \wedge \neg q) \vee r) \Leftrightarrow (\neg(p \vee q) \vee r).$$

Further, by applying the De Morgan's law we obtain

$$\phi \equiv (\neg(p \vee q) \vee r) \Leftrightarrow (\neg(p \vee q) \vee r).$$

Let us put $\psi = (\neg(p \vee q) \vee r)$; thus the analyzed formula takes the form

$$\phi \equiv \psi \Leftrightarrow \psi,$$

which obviously is tautology.

1.6 Normal Forms and Special Forms of Formulae

In this section we recall some important definitions of specific forms of well-formed formulae.

Definition 11. *A literal is any propositional formula p or its negation $\neg p$.*

Literals are the basic components of any formula that is more complex. Two literals, say p and $\neg p$ form the so-called *complementary pair* of literals. A literal without negation will be called *positive*. A literal containing negation will be called *negative*.

1.6.1 Minterms: Simple Conjunctive Formulae

An important concept is the one of a *minterm* [37]; in other words a *simple conjunctive formula*, or a *simple formula*, for short [53]. Such formulae may be used to define a state of a dynamic system or preconditions of a large class of rules.

Definition 12. *Let $q_1, q_2, \ldots, q_n$ be some distinct literals. Any formula of the form*

$$\phi = q_1 \wedge q_2 \wedge \ldots \wedge q_n \tag{1.4}$$

will be called a minterm or a simple formula.

The state (of a certain system) defined with the use of a simple formula is defined in a unique way — there is no use of disjunction and neither of any conditional statements. All the literals — either positive or negative — express certain properties which are either true or false.

There are also two further observations about satisfiability of simple formulae in propositional logic.

Lemma 2. *A minterm (simple conjunctive propositional formula) ϕ is satisfiable iff it does not contain a pair of complementary literals.*

Lemma 3. *A minterm (simple conjunctive propositional formula) ϕ is unsatisfiable iff it contains at least one pair of complementary literals.*

Note, that the use and therefore occurrence of negation sign in formulae is to certain degree a matter of taste and, again to certain degree, can be modified with regard to the current area of application and user's preferences. First, for simplifying the notation and obtaining nice theoretical properties of a certain system one can often avoid the use of the negation symbol ($\neg$) in an explicit way or maybe only positive literals are taken into account[1]. Whenever some kind of negation becomes necessary, implicit, material negation can be used. Instead of writing $\neg$`high_water_level` one can rather write `low_water_level`, instead of $\neg$`switch_on` one can put `switch_off`, etc., i.e. the negation can be often expressed *implicitly*.

More generally, in place of $\neg p$ one can always put a new atom, say np, which is logically equivalent to the negation of p with regard to the *assumed interpretation*. However, note that such an approach may lead to certain problems concerning automated reasoning, if no auxiliary rules (defining all the interdependencies among facts) are defined. For example, no purely logical inference engine will be capable of stating that a formula like `switch_on` $\wedge$ `switch_off` is always false[2] — in such a case an auxiliary reasoning rule like, for example, `switch_on` $\Rightarrow \neg$`switch_off` should be provided so as to assure the detection of inconsistency.

To summarize, any two complementary literals p and $\neg p$ satisfy the following properties:

$$\models p \vee \neg p$$

and

$$\not\models p \wedge \neg p\,.$$

The above properties are recognized at the syntactic level of analysis.

Two literals, say p and q can also be complementary only under specific, assumed interpretation I. In such a case $I(p) = \mathbf{T}$ and $I(q) = \mathbf{F}$ or $I(p) = \mathbf{F}$ and $I(q) = \mathbf{T}$. This can be denoted as:

---

[1] There are many examples of useful systems without explicit negation; the best-known ones are logical AND/OR trees and the Assumption Based Truth Maintenance System (ATMS) of DeKleer [28].

[2] It is false under any admissible interpretation.

$$\models_I p \vee q$$

and

$$\not\models_I p \wedge q\,.$$

The above properties are recognized only at the semantic level of analysis.

A minterm (simple formula) can be considered as a set of its literals. Note that it may be convenient to apply the set notation directly to simple formulae. Let for example ϕ and ψ be two simple formulae, $\phi = p_1 \wedge p_2 \wedge \dots p_k$, $\psi = q_1 \wedge q_2 \wedge \dots q_l$, where both p_i and q_j are literals (either positive or negative ones), $i = 1, 2, \dots, k$, $j = 1, 2, \dots, l$. Then we shall also write $[\phi] = \{p_1, p_2, \dots, p_k\}$ and $[\psi] = \{q_1, q_2, \dots, q_l\}$, and, for example $[\phi] \cup [\psi] = \{p_1, p_2, \dots, p_k\} \cup \{q_1, q_2, \dots, q_l\}$.

Two minterms can be compared to each other — a minterm composed of more literals than the other is *more specific*, while the one containing less literals is *more general*; a more general minterm is satisfied by a larger number of possible interpretations. Consider two satisfiable minterms ϕ and ψ.

Definition 13. *A minterm ϕ subsumes (or is more general than) minterm ψ iff $[\phi] \subseteq [\psi]$.*

Obviously, a more general minterm ϕ is a logical consequence of the more specific one ψ. We have the following lemma.

Lemma 4. *Let ϕ and ψ be two minterms. There is:*

$$\psi \models \phi \qquad \textit{iff} \qquad [\phi] \subseteq [\psi]\,. \tag{1.5}$$

Obviously, logical consequence and subsumption are partial order relations.

1.6.2 Maxterms, Clauses and Rules

A clause is a disjunction of literals. Clauses are used in resolution theorem proving [16, 39]. In propositional calculus a clause is also termed a *maxterm* [37].

Definition 14. *Let $q_1, q_2, \dots, q_n$ be some literals. Any formula of the form*

$$\psi = q_1 \vee q_2 \vee \dots \vee q_n \tag{1.6}$$

will be called a clause or a maxterm.

As in the case of simple formulae, there are also two further observations about satisfiability of clauses in propositional logic.

Lemma 5. *A clause (maxterm) ψ is falsifiable iff it does not contain a pair of complementary literals.*

Lemma 6. *A clause (maxterm) ψ is tautology iff it contains at least one pair of complementary literals.*

Any clause ψ containing at least one positive literal can be transformed into form of a rule (using the symbol of implication). Assume we are given the following clause

$$\psi = \neg p_1 \vee \neg p_2 \vee \ldots \vee \neg p_k \vee h_1 \vee h_2 \vee \ldots \vee h_m \tag{1.7}$$

where $\neg p_1, \neg p_2, \ldots, \neg p_k$ are all the negative literals of ψ. After applying De Morgan's law to the negative literals we obtain

$$\neg(p_1 \wedge p_2 \wedge \ldots \wedge p_k) \vee (h_1 \vee h_2 \vee \ldots \vee h_m)$$

which can be further transformed to equivalent rule form as follows

$$p_1 \wedge p_2 \wedge \ldots \wedge p_k \Rightarrow h_1 \vee h_2 \vee \ldots \vee h_m \,. \tag{1.8}$$

Formula (1.8) constitutes the rule form equivalent to clause (1.7).

In logic, and especially in logic programming, a very important role is played by somewhat restricted form of clauses, i.e. the *Horn clauses.*

Definition 15. *Let $p_1, p_2, \ldots, p_k$ be some positive literals and let h be any literal (either positive or negative). Any formula of the form*

$$\psi = \neg p_1 \vee \neg p_2 \vee \ldots \vee \neg p_k \vee h \tag{1.9}$$

will be called a Horn clause.

A Horn clause is one containing at most one positive literal. According to the above scheme, any Horn clause containing positive literal h can be transformed into rule form, i.e.

$$p_1 \wedge p_2 \wedge \ldots \wedge p_k \Rightarrow h \,. \tag{1.10}$$

Formula (1.10) constitutes the rule form equivalent to clause (1.9). Such rules constitute an important form for knowledge representation — they constitute the core of logic programming and rule-based systems.

Any clause can be considered as a set of its literals. Then it may be convenient to apply the set notation directly to its elements. Let for example ψ be a clause, $\psi = p_1 \vee p_2 \vee \ldots p_k$, where p_i are literals (either positive or negative), $i = 1, 2, \ldots, k$. In order to denote the set of literals of a clause we shall write $[\psi] = \{p_1, p_2, \ldots, p_k\}$.

Two clauses can be compared to each other — a clause composed of more literals than the other is *more general,* while the one containing less literals is *more specific*; a more general clause is satisfied by a larger number of possible interpretations. Consider two falsifiable clauses ϕ and ψ.

Definition 16. *A clause ψ subsumes (or is more specific than) clause φ iff* $[\psi] \subseteq [\varphi]$.

Obviously, a more general clause φ is a logical consequence of the more specific one ψ. We have the following lemma.

Lemma 7. *Let ψ and φ be two clauses. There is:*

$$\psi \models \varphi \qquad \textit{iff} \qquad [\psi] \subseteq [\varphi]\,. \tag{1.11}$$

Obviously, both logical consequence and subsumption are partial order relations.

1.6.3 Conjunctive Normal Form

Conjunctive Normal Form (CNF) is a form having the structure of a conjunction of clauses. It is very regular and thus transparent. It may be used in automated theorem proving with resolution.

Definition 17 (CNF). *A formula Ψ is in Conjunctive Normal Form (CNF) if it can be presented as*

$$\Psi = \psi_1 \wedge \psi_2 \wedge \ldots \wedge \psi_n \tag{1.12}$$

where $\psi_1, \psi_2, \ldots, \psi_n$ denote any clauses.

Thus in fact any formula in CNF constitutes a two-level structure: at the first level one has a set of some clauses while at the second level the clauses are connected with conjunction.

Any formula in CNF can be considered as a set of its clauses. Then it may be convenient to apply the set notation directly to its elements. Let for example Ψ be a formula in CNF, $\Psi = \psi_1 \wedge \psi_2 \wedge \ldots \wedge \psi_n$. In order to denote the set of clauses we shall write $[\Psi] = \{\psi_1, \psi_2, \ldots, \psi_n\}$.

Basing on the structure of CNF the following simple observation can be put forward. Since the formula is a conjunction of clauses, in order to demonstrate its unsatisfiability it is enough to find a subset of these clauses which is unsatisfiable. Thus, roughly speaking, when attempting at proving unsatisfiability of a formula it seems reasonable to transform it into an appropriate CNF. In an extreme case it may contain a conjunction of complementary literals, which would make the proof straightforward.

As may be observed, for a particular formula there may exist many different CNF which are equivalent. In fact, the CNF for an arbitrary formula is not defined in a unique way. However, it is possible to choose one specific form which is unique; it is the CNF composed of maximal clauses, i.e. ones composed of all the propositional symbols occurring in the initial formula.

Definition 18. *Let Ψ denote a well-formed propositional formula of arbitrary structure and let P_Ψ denote the set of all propositional symbols the formula is built with. A maximal clause ψ is one composed of all the symbols of P_Ψ (either negated or positive). A maximal CNF of Ψ is the formula defined as*

$$maxCNF(\Psi) = \psi_1 \wedge \psi_2 \wedge \ldots \wedge \psi_n \tag{1.13}$$

where all the clauses $\psi_1, \psi_2, \ldots, \psi_n$ are maximal.

The maximal CNF form is also known as *full conjunctive normal form* or *conjunctive canonical form.*

Maximal CNF of a formula can be obtained through transformation of the initial formula to CNF; then, any clause ψ which is not a maximal one, i.e. it lacks some propositional variable q, should be extended according to the following scheme

$$\psi \quad \longrightarrow \quad \psi \vee (q \wedge \neg q) \quad \longrightarrow \quad (\psi \vee q) \wedge (\psi \vee \neg q) \,.$$

In this way any clause can be extended to contain any missing propositional symbol.

For technical applications it is useful to describe also minimal form of a formula in CNF; such minimal forms are usually the base for technical implementations.

Definition 19. *A formula*

$$\Psi = \psi_1 \wedge \psi_2 \wedge \ldots \wedge \psi_n$$

is in minimal CNF form iff there does not exist a logically equivalent formula in CNF composed of m minterms where $m < n$.

It can be noticed that in general case the minimal CNF of a formula is not defined in a unique way. The problem of minimal CNF and formulae minimization is important for technical applications, since the number of elements necessary to implement an appropriate circuit is maximally reduced. Some methods of finding minimal forms with the so-called Karnaugh tables are presented in [115].

1.6.4 Disjunctive Normal Form

Disjunctive Normal Form (DNF) is a form having the structure of a disjunction of simple formulae. It is also very regular and thus transparent. It may be used in automated theorem proving with dual resolution [53, 54, 55, 56], in analysis of rule-based systems [57, 60] and for modeling states of dynamical systems [53, 57].

Definition 20 (DNF). *A formula Φ is in Disjunctive Normal Form (DNF) if it can be presented as*

$$\Phi = \phi_1 \vee \phi_2 \vee \ldots \vee \phi_n \qquad (1.14)$$

where $\phi_1, \phi_2, \ldots, \phi_n$ denote any minterms (simple formulae).

Thus in fact any formula in DNF constitutes a two-level structure: at the first level one has a set of some simple conjunctive formulae while at the second level these formulae are connected with disjunction.

Any formula in DNF can be considered as a set of its minterms. Then it may be convenient to apply the set notation directly to its elements. Let for example Φ be a formula in DNF, $\Phi = \phi_1 \vee \phi_2 \vee \ldots \vee \phi_n$. In order to denote the set of all the minterms of Φ we shall write $[\Phi] = \{\phi_1, \phi_2, \ldots, \phi_n\}$.

Basing on the structure of DNF the following simple observation can be put forward. Since the formula is a disjunction of conjunctive formulae, in order to demonstrate its satisfiability it is enough to find a single minterm component which is satisfiable. Further, in order to show that the formula is tautology it is enough to find a subset of its minterm components which form tautology. Thus, roughly speaking, when attempting to prove the validity of a formula it seems reasonable to transform it into an appropriate DNF. In an extreme case it may contain a disjunction of complementary literals, which would make the proof straightforward.

As may be observed, for a particular formula there may exist many different DNF equivalents. In fact, the DNF for an arbitrary formula is not defined in a unique way. However, it is possible to choose the one specific form which is unique; it is the DNF composed of maximal minterms, i.e. ones composed of all the propositional symbols occurring in the initial formula.

Definition 21. *Let Φ denote a well-formed propositional formula of arbitrary structure and let P_Φ denote the set of all propositional symbols the formula is build with. A maximal minterm ϕ is one composed of all the symbols of P_Φ (either negated or positive). A maximal DNF of Ψ is the formula defined as*

$$\max \mathrm{DNF}(\Psi) = \psi_1 \vee \psi_2 \vee \ldots \vee \psi_n \qquad (1.15)$$

where all the minterms $\psi_1, \psi_2, \ldots, \psi_n$ are maximal.

The maximal DNF form is also known as *full disjunctive normal form* or *disjunctive canonical form*.

Maximal DNF of a formula can be obtained through transformation of the initial formula to DNF; then, any minterm ψ which is not a maximal one, i.e. it lacks some propositional variable q, should be extended according to the following scheme

$$\psi \quad \longrightarrow \quad \psi \wedge (q \vee \neg q) \quad \longrightarrow \quad \psi \wedge q \vee \psi \wedge \neg q\,.$$

In this way any minterm can be extended to contain any missing propositional symbol.

For technical applications it is useful to describe also minimal form of a formula in DNF; such minimal forms are usually the base for technical implementations.

Definition 22. *A formula*

$$\Phi = \phi_1 \vee \phi_2 \vee \ldots \vee \phi_n$$

is in minimal DNF form iff there does not exist a logically equivalent formula in DNF composed of m minterms where $m < n$.

It can be noticed that in general case the minimal DNF of a formula is not defined in a unique way. The problem of minimal DNF and formulae minimization will be discussed with respect to rule-based systems reduction. We shall return to finding minimal forms of rule-based systems through gluing rules in chapter 15 after introducing the so-called *dual resolution method.*

1.6.5 Transformation of a Formula into CNF/DNF

Any well-formed formula can be transformed into a logically equivalent CNF or DNF. This is achieved by subsequent application of transformation rules given in Sect. 1.4.

In order to transform any formula into an equivalent CNF or DNF the following steps should be carried out:

1. $\Phi \Leftrightarrow \Psi \equiv (\Phi \Rightarrow \Psi) \wedge (\Psi \Rightarrow \Phi)$ — elimination of equivalence symbols,
2. $\Phi \Rightarrow \Psi \equiv \neg\Phi \vee \Psi$ — elimination of implications,
3. $\neg(\neg\Phi) \equiv \Phi$ — elimination of nested negations,
4. $\neg(\Phi \vee \Psi) \equiv \neg\Phi \wedge \neg\Psi$ — application of De Morgan's law to move the negation sign directly to propositional symbols,
5. $\neg(\Phi \wedge \Psi) \equiv \neg\Phi \vee \neg\Psi$ — application of De Morgan's law to move the negation sign directly to propositional symbols,
6. $\Phi \vee (\Psi \wedge \Upsilon) \equiv (\Phi \vee \Psi) \wedge (\Phi \vee \Upsilon)$ — application of distributivity law for transforming to CNF,
7. $\Phi \wedge (\Psi \vee \Upsilon) \equiv (\Phi \wedge \Psi) \vee (\Phi \wedge \Upsilon)$ — application of distributivity law for transforming to DNF.

Example. The following formula will be transformed to DNF.

$$\begin{aligned}(p \wedge (p \Rightarrow q)) \Rightarrow q &= \neg(p \wedge (p \Rightarrow q)) \Rightarrow q = \neg(p \wedge \neg(p \vee q)) \vee q \\ &= (\neg p \vee \neg(\neg p \vee q)) \vee q = (\neg p \vee p \wedge \neg q) \vee q \\ &= (\neg p \wedge q \vee \neg p \wedge \neg q \vee p \wedge \neg q \vee q \wedge p \vee q \wedge \neg p) \\ &= \neg p \wedge (q \vee \neg q) \vee p \wedge (q \vee \neg q) = \neg p \vee p = \top\,.\end{aligned}$$

The formula is tautology; the obtained DNF is $\top$.

1.6.6 Example

In this section let us consider a simple but complete example of formula analysis; the initial version of the problem and its analysis comes from [115], although we carry out a more complete analysis here. For given two formulae we shall analyze the possibility of checking if one of them is the logical consequence of the other. We shall apply the tabular method based on checking of all possible interpretations and transformation to DNF, maximal DNF, and minimal DNF.

Consider the following two formulae:

$$\phi = (p \Rightarrow q) \wedge (r \Rightarrow s)$$

and

$$\varphi = (p \vee r) \Rightarrow (q \vee s) \, .$$

The basic task is to check whether

$$\phi \models \varphi. \tag{1.16}$$

Since there are as many as 4 propositional symbols, there exist 16 possible interpretations. Let us examine the first approach based on extensive checking of all the interpretations. We obtain the matrix given by (Table 1.2).

After short analysis of the columns representing ϕ and φ it is obvious that in fact $\phi \models \varphi$.

Table 1.2. Logical matrix for checking if $\phi \models \varphi$, where $\phi = (p \Rightarrow q) \wedge (r \Rightarrow s)$ and $\varphi = (p \vee r) \Rightarrow (q \vee s)$

p	q	r	s	$p \Rightarrow q$	$r \Rightarrow s$	ϕ	$p \vee r$	$q \vee s$	φ
0	0	0	0	1	1	1	0	0	1
0	0	0	1	1	1	1	0	1	1
0	0	1	0	1	0	0	1	0	0
0	0	1	1	1	1	1	1	1	1
0	1	0	0	1	1	1	0	1	1
0	1	0	1	1	1	1	0	1	1
0	1	1	0	1	0	0	1	1	1
0	1	1	1	1	1	1	1	1	1
1	0	0	0	0	1	0	1	0	0
1	0	0	1	0	1	0	1	1	1
1	0	1	0	0	0	0	1	0	0
1	0	1	1	0	1	0	1	1	1
1	1	0	0	1	1	1	1	1	1
1	1	0	1	1	1	1	1	1	1
1	1	1	0	1	0	0	1	1	1
1	1	1	1	1	1	1	1	1	1

Now, let us transform ϕ to DNF; we obtain:

$$\begin{aligned}\phi &= (p \Rightarrow q) \wedge (r \Rightarrow s) = (\neg p \vee q) \wedge (\neg r \vee s) \\ &= (\neg p \wedge \neg r) \vee (\neg p \wedge s) \vee (q \wedge \neg r) \vee (q \wedge s) \ .\end{aligned}$$

In this way we have obtained a DNF for ϕ. One can check that it is also the minimal form of DNF for ϕ. The maximal form is given by:

$$\begin{aligned}\max \mathrm{DNF}(\phi) = &(\neg p \wedge \neg q \wedge \neg r \wedge \neg s) \vee (\neg p \wedge \neg q \wedge \neg r \wedge s) \vee (\neg p \wedge \neg q \wedge r \wedge s) \vee \\ &(\neg p \wedge q \wedge \neg r \wedge \neg s) \vee (\neg p \wedge q \wedge \neg r \wedge s) \vee (\neg p \wedge q \wedge r \wedge s) \vee \\ &(p \wedge q \wedge \neg r \wedge \neg s) \vee (p \wedge q \wedge \neg r \wedge s) \vee (p \wedge q \wedge r \wedge s) \ .\end{aligned}$$

Now, let us transform φ to DNF; we obtain:

$$\begin{aligned}\phi &= (p \vee r) \Rightarrow (q \vee s) = \neg(p \vee r) \vee q \vee s = (\neg p \wedge \neg r) \vee q \vee s \\ &= (\neg p \wedge \neg r) \vee q \vee s \ .\end{aligned}$$

In this way we have obtained a DNF for φ. One can check that it is also the minimal form of DNF for φ. The maximal form is given by:

$$\begin{aligned}\max \mathrm{DNF}(\phi) = &(\neg p \wedge \neg q \wedge \neg r \wedge \neg s) \vee (\neg p \wedge \neg q \wedge \neg r \wedge s) \vee (\neg p \wedge \neg q \wedge r \wedge s) \vee \\ &(\neg p \wedge q \wedge \neg r \wedge \neg s) \vee (\neg p \wedge q \wedge \neg r \wedge s) \vee (\neg p \wedge q \wedge r \wedge s) \vee \\ &(\neg p \wedge q \wedge r \wedge \neg s) \vee (p \wedge q \wedge \neg r \wedge \neg s) \vee (p \wedge q \wedge \neg r \wedge s) \vee \\ &(p \wedge q \wedge r \wedge s) \vee (p \wedge q \wedge r \wedge \neg s) \vee (p \wedge \neg q \wedge \neg r \wedge s) \vee \\ &(p \wedge \neg q \wedge r \wedge s) \ .\end{aligned}$$

As it can be observed, due to uniqueness of the maximal DNF for any formula, checking for logical entailment of these forms is straightforward. The necessary and sufficient condition is that

$$[\max \mathrm{DNF}(\phi)] \subseteq [\max \mathrm{DNF}(\varphi)] \ ,$$

i.e. all the maximal minterms of the more general formula must occur also in the less general one.

This result suggests a simple method for checking if logical entailment holds; unfortunately, this kind of check is computationally costly. In case of more complex formulae the maximal DNF may appear to be very large. In further sections a more efficient approach will be put forward.

1.7 Logical Consequence and Deduction

Let us recall the notion of *logical consequence* formally specified in Definition 9. Formula φ is a logical consequence of formula ϕ iff φ is satisfied under any interpretation satisfying ϕ ($\phi \models \varphi$).

Note that the relation of logical consequence is a partial order relation, i.e. it is reflexive, antisymmetric and transitive [53]. The relation of logical equivalence is an equivalence relation [53].

The concept of logical consequence is of primary importance in the whole logic. One may say, that one of the main issues of interest in logic are methods and rules for analysis of the relationship of logical consequence.

The most typical situation is as follows. There is specified certain knowledge about a system under consideration, formalized as a set of assumptions and observations, to be denoted by Δ. There is also some statement of interest H, i.e. a hypothesis formalizing some further knowledge about the world. The main problem of interest is to check if

$$\Delta \models H \,, \tag{1.17}$$

i.e. to check if the hypothesis logically follows from the accepted knowledge. The process of showing that H is a valid consequence of Δ is called *theorem proving*.

There are two different approaches to theorem proving originating from separate ideas. The first one is based on direct checking of all the possible interpretations. In fact, it is enough to check only those under which Δ is true; for any such interpretation H must be true as well, so as to be a logical consequence of the assumptions. We have shown an example application of this approach in the first part of Sect. 1.6.6.

In the second approach the check is accomplished by applying some formal rules (productions) to *derive* some or all of the formulae which are logical consequences of Δ. A crucial issue here is that any derivation rule applied must be *sound*, i.e. it must produce a formula being logical consequence of its predecessor formulae.

Both of the approaches are potentially applicable, and both suffer from combinatorial explosion. One can estimate that, in case of direct checking of possible interpretations, in case of n propositional variables, there are as many as 2^n potential interpretations. This means that practical use of the first approach is limited to formulae of some reasonably small dimension.

In case of the other approach, the combinatorial explosion occurs since there are (i) many inference rules and, further, (ii) they can be applied in many different ways each, what leads to producing many new formulae at every stage of inference. The graph representing schematically the course of inference is usually characterized by rapidly exploding number of nodes representing newly generated formulae.

Finally, the check for logical entailment specified by formula (1.17) can be rendered to the problem of proving that certain formula is valid (or inconsistent). This is so thanks to the *Deduction Theorems* [16].

Theorem 1 (Deduction I). *Formula H is a logical consequence of Δ iff formula $\Delta \Rightarrow H$ is tautology; this is written shortly as:*

$$\Delta \models H \quad \text{iff} \quad \models (\Delta \Rightarrow H) \,. \tag{1.18}$$

Theorem 2 (Deduction II). *Formula H is a logical consequence of Δ iff formula $\Delta \wedge \neg H$ is unsatisfiable; this is written shortly as:*

$$\Delta \models H \qquad \textit{iff} \qquad \not\models (\Delta \wedge \neg H)\,. \tag{1.19}$$

The latter of the above theorems is frequently used in automated theorem proving, especially ones based on the Resolution method.

1.8 Inference Modes: Deduction, Abduction and Induction

In this section we refer to reasoning, logical inference, and especially various modes of reasoning and inference rules.

There are a number of modes of reasoning; people apply perhaps more than twenty somewhat different modes. Some of the modes are 'logically correct', some are plausible or heuristic, some are more akin to guesswork; the reasoning can be precise or vague, probabilistic, fuzzy, rough, uncertain, etc.

Some best known, named reasoning modes include the following:

- Generalization — reasoning from details to general statements.
- Specialization — reasoning from general statements to details.
- Classification — through pattern matching, similarity or distance analysis, etc.
- Classical logical inference:
  - deduction,
  - abduction,
  - induction.
- Consistency-based reasoning.
- Reasoning through analogy, case-based reasoning.
- Non-monotonic reasoning.
- Probabilistic reasoning.
- Heuristic reasoning, plausible reasoning.
- Imprecise reasoning, fuzzy reasoning, rough reasoning.
- Parametric reasoning, application of specialized numerical procedures (Neural Networks, optimization, adaptation, etc.).
- Reasoning through narrowing.
- Reasoning through excluding.
- Application of search algorithms.

In logic the purely logical reasoning paradigms are in point of interest; these are: deduction, abduction, and induction. The most important, and perhaps the best known is deduction, as it is studied in practically all logic handbooks and lectured on every course on logic [36, 39, 125].

Reasoning becomes 'inference' if it is performed according to a specific scheme. Note that the term 'inference' or more precisely 'logical inference' is quite a wide one. Inference is performed by generating new statements or formulae from a certain initial set of assumptions and with the use of specific set of inference rules. Inference rules define various schemes of production of new formulae from ones assumed to form initial knowledge base or ones that have been derived earlier.

The generic scheme for any inference (logical inference, if logical reasoning rules are applied) is as follows. There is a set of initial knowledge items (formulae), say Δ, and a set of inference rules R. Every rule $r \in R$ consists of two parts: the *premise* or *precondition* α and the *conclusion* or *thesis* β. Both the premise and the conclusion can be composed of a number of items, i.e. $\alpha = \{\alpha_1, \alpha_2, \ldots, \alpha_m\}$ and $\beta = \{\beta_1, \beta_2, \ldots, \beta_n\}$. A rule like that is usually denoted as:

$$\alpha \longrightarrow \beta,$$

i.e.

$$\alpha_1, \alpha_2, \ldots, \alpha_m \longrightarrow \beta_1, \beta_2, \ldots, \beta_n$$

or in the vertical form as

$$\frac{\alpha_1, \alpha_2, \ldots, \alpha_m}{\beta_1, \beta_2, \ldots, \beta_n}.$$

Let Σ denote the current set describing the generated knowledge; in the beginning one puts $\Sigma = \Delta$. A single step of inference consists of:

- selecting a rule $r \in R$,
- checking if the rule is applicable to Σ,
- if so, generating the new knowledge Σ'.

A rule is applicable if its conditions are satisfied; a basic check may be accomplished by verifying if $\alpha_1, \alpha_2, \ldots, \alpha_m \in \Sigma$ although in more complex logics proof procedures may be required[3]. If so, the rule is applicable and new knowledge is generated as $\Sigma' = \Sigma \cup \{\beta_1, \beta_2, \ldots, \beta_n\}$.

In case there exists a sequence of rules, such that after successive application to Δ some Σ was obtained, we say that Σ is derived from Δ; this is written shortly as

$$\Delta \vdash \Sigma,$$

or $\Delta \vdash_R \Sigma$ in case it is intended to underline that Σ was derived with a specific set of rules. The process is called *derivation* (logical derivation).

1.8.1 Deduction Rules for Propositional Logic

Deduction [37, 39, 115, 125] is the best known method of logical inference. Roughly speaking, deduction is performed by generation of new valid statements from some initial statements known to be valid with use of valid inference rules. The basic property of deduction is that from valid knowledge

[3] Some more advanced checks may require substituting for variables based on unification or resolution theorem proving.

only valid statements are obtained. Thus, the wide range of applications of deduction contains tasks such as as theorem proving, hypotheses verification, knowledge generation, question answering systems, deductive databases, etc.

As an example of a deductive rule consider the well-known *modus ponens scheme*

$$\frac{\alpha, \alpha \Rightarrow \beta}{\beta},$$

where α and β are arbitrary formulae. This inference rule seems to be intuitively obvious, constitutes the most popular deductive inference paradigm. Obviously, it is sound, but unfortunately incomplete — not all true statements can be generated from the assumed set of axioms. To see that, consider another sound inference paradigm, known as *transition rule*

$$\frac{\alpha \Rightarrow \beta, \beta \Rightarrow \gamma}{\alpha \Rightarrow \gamma}.$$

Obviously, this inference paradigm is also valid, however, the final conclusion cannot be obtained with the previous rule.

The general scheme of deductive inference can be codified as follows

$$\frac{A_1, A_2, \ldots, A_m, R}{B_1, B_2, \ldots, B_n},$$

where $A_1, A_2, \ldots, A_m$ are prerequisites, axioms, earlier generated facts or formulae known to be valid, R is a set of inference rules, and $B_1, B_2, \ldots, B_n$ are conclusions, such that

$$A_1, A_2, \ldots, A_m, R \models B_1, B_2, \ldots, B_n .$$

What is important and characteristic for deduction rules and derivation based on deduction is that whenever

$$\varDelta \vdash \varSigma$$

then there is also

$$\varDelta \models \varSigma .$$

In other words, deduction is a valid inference procedure — from correct assumptions only correct conclusions are derived.

A number of deduction rules are specified in literature. In automated theorem proving and AI the method of *resolution* [16, 39] or [37] gained a lot of attention, for being based on a single and powerful rule, relatively easy to implement.

Some of the most important deduction rules are the following:

- $\dfrac{\alpha}{\alpha \vee \beta}$ — disjunction introduction,
- $\dfrac{\alpha,\ \beta}{\alpha \wedge \beta}$ — conjunction introduction,

- $\dfrac{\alpha \wedge \beta}{\alpha}$ — conjunction deletion,
- $\dfrac{\alpha,\ \alpha \Rightarrow \beta}{\beta}$ — modus ponens (modus ponendo ponens),
- $\dfrac{\alpha \Rightarrow \beta,\ \neg\beta}{\neg\alpha}$ — modus tollens (modus tollendo tollens),
- $\dfrac{\alpha \vee \beta,\ \neg\alpha}{\beta}$ — modus tollendo ponens,
- $\dfrac{\alpha \bigoplus \beta,\ \alpha}{\neg\beta}$ — modus ponendo tollens,
- $\dfrac{\alpha \Rightarrow \beta,\ \beta \Rightarrow \gamma}{\alpha \Rightarrow \gamma}$ — transitivity rule,
- $\dfrac{\alpha \vee \gamma,\ \neg\gamma \vee \beta}{\alpha \vee \beta}$ — resolution rule,
- $\dfrac{\alpha \wedge \gamma,\ \neg\gamma \wedge \beta}{\alpha \wedge \beta}$ — (backward) dual resolution (works backwards),
- $\dfrac{\alpha \Rightarrow \beta,\ \gamma \Rightarrow \delta}{(\alpha \vee \gamma) \Rightarrow (\beta \vee \gamma)}$ — constructive dilemma law,
- $\dfrac{\alpha \Rightarrow \beta,\ \gamma \Rightarrow \delta}{(\alpha \wedge \gamma) \Rightarrow (\beta \wedge \delta)}$ — constructive dilemma law.

There are many other deduction rules, however, for the rest of the material two rules are important. These are *resolution* and *dual resolution*. Both the rules provide powerful inference paradigms.

1.8.2 Resolution Rule

Resolution rule constitutes a single and powerful inference rule of conceptually simple scheme. It is very attractive, especially for automated theorem proving. Below we explain a single step in resolution theorem proving, i.e. the *resolution rule and its application*.

Let there be given two propositional clauses, $C_1 = \phi \vee q$ and $C_2 = \varphi \vee \neg q$. It is important that there is a pair of complementary literals q and $\neg q$, occurring in the clauses. The resolution rule (or resolution principle) allows to generate a new clause $C = \phi \vee \varphi$ being a logical consequence of the parent clauses; the complementary literals are removed.

Definition 23. *Let $C_1 = \phi \vee q$ and $C_2 = \varphi \vee \neg q$ be two arbitrary clauses. The Resolution Rule is an inference rule of the form*

$$\frac{\phi \vee q, \varphi \vee \neg q}{\phi \vee \varphi}\,. \tag{1.20}$$

Obviously, the produced formula is a logical consequence of the parent formulae. Further, note that at the level of propositional language the resolution rule is equivalent to the rule expressing transitivity. In order to see that let us transform the clauses to the following equivalent forms, i.e. $C_1 = \neg\phi \Rightarrow q$ and $C_2 = q \Rightarrow \varphi$. Now, the resolution rule takes the form

$$\frac{\neg\phi \Rightarrow q, q \Rightarrow \varphi}{\neg\phi \Rightarrow \varphi}.$$

The resulting formula can be further transformed so we have $\neg\phi \Rightarrow \varphi \equiv \phi \vee \varphi$.

Resolution theorem proving consists of repeated steps of appropriate application of the resolution rule. As it was mentioned, the method is especially convenient for automated theorem proving. It has gained a great popularity during the last thirty years. The resolution method ([16, 36, 37, 39], for precise, logical treatment see as well [144]) combines in a single rule the power of other rules, and due to its uniformity, can be easily implemented for automated theorem proving with computers.

Let us briefly recapitulate some essence of the resolution method. As presented in [16, 39], the idea of resolution theorem proving lies in *refutation* of the negated formula to be proved. Without going into the very details, resolution theorem proving proceeds as follows.

Let Δ denote a set of given axioms (from logical point of view, a *conjunction* of them), and let H be a formula (e.g. a hypothesis) to be proved. Thus, according to the Deduction Theorem (1) one is to prove that

$$\Delta \Rightarrow H \tag{1.21}$$

is a valid formula. In classical resolution method, instead of proving (1.21), one takes the negation of it, i.e.

$$\Delta \wedge \neg H \tag{1.22}$$

and tries to show that (1.22) is unsatisfiable; this approach is based on the Deduction Theorem (2). Then, (1.22) is transformed into CNF, i.e. to a set (conjunction) of clauses (disjunctions of literals). Now, in order to show that a set of clauses is unsatisfiable, one attempts to derive an empty clause $\perp$ (here: always unsatisfiable) from it. The derivation is carried out with the use of resolution rule, which preserves logical consequence. Thus, any newly derived clause is a logical consequence of its parent clauses. If an empty formula is eventually derived, the unsatisfiability of the initial set of clauses is proved.

A great advantage of the classical resolution method lies in leaving the set of axioms Δ almost unchanged. In most of practical cases there is a set (conjunction) of separate axioms, and each of them can be converted into clausal form independently from the other. This approach saves computational effort and fits well into most of classical problems in a naturally efficient way. Together with simplicity of the resolution rule, this constitutes probably the reason for the tremendous success and popularity of resolution theorem proving. However, from theoretical point of view, one can also try to prove validity of (1.21) directly. Below, we present the basic idea for such an approach using the idea of bd-resolution.

1.8.3 Dual Resolution Rule

Dual Resolution Method is a method of automated theorem proving analogous to classical resolution; however, instead of starting from CNF and showing its unsatisfiability, dual resolution starts with DNF and attempts at demonstrating that this formula is valid. The method is especially convenient for proving generalization and completeness rather than automated theorem proving in general; moreover, theorem proving as such is not a matter of primary interest and importance in this work. However, it may be worthwhile to further analyze the possibility of applying the method for general theorem proving, since such a discussion seems to be ignored in literature[4].

Since the method can be regarded as analogical (dual) to the classical resolution method, the presentation refers to resolution theorem proving as a kind of a 'reference point'. A basic comparative analysis of these two methods will be carried out as well.

The bd-resolution presented in this work is analogous to resolution method. The main difference is that instead of checking *unsatisfiability* of a set of clauses we rather try to prove *validity* of the given initial formula transformed into disjunctive normal form. Thus, the initial form is in fact *dual* to the one used in resolution method.

Further, the proposed method works in fact *backwards*. This means that during the process of derivation one generates new formulae from parent formulae starting from the initial formula to be proved — but with regard to logical inference, it is the disjunction of parent formulae which is a logical consequence of the derived formula! In other words, during generation of the proof, one is to search for premises from which the formula to be proved follows. Therefore, at any step of reasoning the derivation process is reversed with regard to finding logical consequences. The process of derivation is successful if it eventually ends up with an empty formula $\top$, which (here) is always true — in this case, the initial formula, as a logical consequence of it, is proved.

As resolution rule, its dual version constitutes a single and powerful inference rule of conceptually simple scheme. It is as simple as resolution, but the preferable area of applications is different[5]. Below we explain a single

[4] A short note on the *consolution* method, which in fact is logically equivalent to dual resolution, is given in [8]; some discussion is given also in [7], however, it does not explain why dual resolution is not used. A more detailed discussion of possibility of application of dual resolution method to theorem proving is given in [53].

[5] In [7] one can find a short discussion why resolution is based on refutation and not on affirmative theorem proving; one of the conclusions is that it is more convenient

step of dual resolution theorem proving, i.e. the *dual resolution rule* and its application.

Let there be given two propositional minterms, $M_1 = \phi \wedge q$ and $M_2 = \varphi \wedge \neg q$. It is important that there is a pair of complementary literals q and $\neg q$, occurring in the minterms. The dual resolution rule (or dual resolution principle) allows to generate a new minterm $M = \phi \wedge \varphi$ such that the disjunction of the parent clauses is a logical consequence of it; the complementary literals are removed.

Definition 24. *Let $M_1 = \phi \wedge q$ and $M_2 = \varphi \wedge \neg q$ be two arbitrary clauses. The Dual Resolution Rule is an inference rule of the form*

$$\frac{\phi \wedge \ q, \varphi \wedge \neg q}{\phi \wedge \varphi} . \tag{1.23}$$

Obviously, the produced formula is not a logical consequence of the parent formulae! It is the disjunction of the parent formulae which is logical consequence of the resulting formula. Thus the logical consequence goes in fact *backwards* with respect to derivation direction. Let $\vdash_{DR}$ stay for 'derived with dual resolution'. We shall prove the following theorem.

Theorem 3. *If*

$$M_1 \vee M_2 \vdash_{DR} M \ ,$$

then

$$M \models M_1 \vee M_2 \ .$$

Proof. Assume $M = \phi \wedge \varphi$ is satisfied under some arbitrary interpretation I. For two complementary literals q, there must be either $\models_I q$ or $\models_I \neg q$. If the first possibility is true, then $M_1 = \phi \wedge q$ is satisfied, and so is $M_1 \vee M_2$. In the other case, $M_2 = \varphi \wedge \neg q$ must be satisfied, and so must be $M_1 \vee M_2$. □

With respect to Theorem 3, the logical consequence shows *backwards*, and thus the dual resolution rule is also termed *backward dual resolution* or *bd-resolution*, for short [53–56]

Theorem proving with dual resolution is performed by appropriate application of the dual resolution rule. As it was mentioned, the method is especially convenient for completeness checking and proving that a DNF formula is more general than another one, but it can also be applied for automated theorem proving.

---

and follows from tradition. In fact, as pointed in Sect 1.8.2, the advantage is that resolution allows to transform any of the axioms to clausal form separately, leaving the structure of the initial conjunction. However, contrary to Bibel [7] we would not call affirmative calculus a 'heresy'; as it can be seen in this book, there are areas of applications, where dual resolution seems to be a reasonable approach — such an area is, for example, verification of completeness of a set of rules.

Note that instead of converting a formula to its clausal form one can transform the formula to a dual form, consisting of a disjunction of minterms, i.e. the DNF. After negating the clausal form (i.e. the form used in resolution theorem proving[6] of (1.22) and successful application of De Morgan's laws, the dual form is obtained. Of course, it would be more reasonable to directly convert (1.21) into the appropriate form.[7]

For further considerations we assume that the formula to be proved valid (1.21) is in DNF. Thus, the formula can be written as

$$\Phi = \phi_1 \vee \phi_2 \vee \ldots \vee \phi_n \,. \tag{1.24}$$

Now, we can state the following theorem.

Theorem 4. *Let Φ be a DNF formula defined by* (1.24)*. If an empty formula $\top$ (always true) can be derived from Φ with dual resolution rule, then Φ is valid (is a tautology).*

Proof. By Theorem 3, for any simple formula ϕ derived from Φ there is $\phi \models \Phi$.

Consider some middle step of inference. We have the current formula Φ_i, and ϕ_i derived from it by dual resolution. For the next step we use $\Phi_{i+1} = \Phi_i \vee \phi_i$, and the next dual resolvent is ϕ_{i+1}, and the process goes on like that.

Obviously, by Theorem 3 $\phi_i \models \Phi_i$. Further, $\phi_{i+1} \models \Phi_{i+1}$. Assume I is any interpretation such that $\models_I \phi_{i+1}$; thus $\models_I \Phi_{i+1}$, i.e. $\models_I \Phi_i \vee \phi_i$. Now, if $\models_I \phi_i$, then due to transitivity of logical entailment $\models_I \Phi_i$; if not, we have directly $\models_I \Phi_i$. Hence $\phi_{i+1} \models \Phi_i$. By induction with respect to the length of the derivation we have $\phi_k \models \Phi$ for any k. To conclude, if $\phi_k = \top$, then the initial formula Φ is valid. □

Now, in order to illustrate the possibility of theorem proving with the use of bd-resolution let us present a simple example.

Example. We shall demonstrate the use of bd-resolution on a simple example in order to provide some intuitions. Let us take the so-called *implication distribution scheme* (see, for example [39], p. 56). Using bd-resolution we shall prove that implication distribution scheme for propositional formulae is in fact a tautology. The analyzed scheme is as follows:

$$(p \Rightarrow (q \Rightarrow r)) \Rightarrow ((p \Rightarrow q) \Rightarrow (p \Rightarrow r)) \,. \tag{1.25}$$

After transformation to DNF, one obtains from (1.25) a disjunction of four simple formulae (see below, formulae 1 to 4); the derivation of an empty formula (always true) can proceed as follows:

---

[6] A method of transforming any formula to such a form is described in e.g. [16] or [39]).

[7] Such a transformation, analogous to the one applied in classical resolution, is described in [7]; during the transformation Skolemization is used to remove *universal* quantification [53].

1. $p \wedge q \wedge \neg r$, (from (1.25))
2. $p \wedge \neg q$, (from (1.25))
3. $\neg p$, (from (1.25))
4. r, (from (1.25))
5. $p \wedge q$, (from 1. and 4.)
6. $\neg q$, (from 2. and 3.)
7. p, (from 5. and 6.)
8. $\top$. (from 3. and 7.)

Thus, the initial formula (1.25) is valid. From this example one can see that bd-resolution can be applied for efficient proving of formulae which are easily transformed into disjunctive normal form.

1.9 Abduction and Induction

Apart from deduction, there are two other logical inference modes. These are *abduction* and *induction*. They are not so popular as deduction and they do not have the very basic property of deduction — they are not valid inference paradigms. This means that the produced output rules are not necessarily logical consequences of the input ones, and as such they constitute rather some hypotheses than unconditionally true statements. However, despite the problem of validity, they are interesting and useful schemes of reasoning.

1.9.1 Abduction

Abduction is another method of logical inference, however, many people do not classify it as a strictly logical rule. The reason for that is the following: abduction generates statements which are not necessarily true. In fact, abduction provides *hypotheses* only, i.e. statements which are possible, potential solutions, usually non-unique, and not necessarily correct.

In order to explain the basic idea of abduction it is enough to say that abduction looks for certain assumption under which the specified conclusions are true with respect to some known domain theory.

A simple example of abductive rule may be the following statement

$$\frac{\alpha \Rightarrow \beta, \beta}{\alpha}.$$

It is to be read: if α implies β and β is true, then α is true. Obviously, abduction is not a valid inference rule; in fact it generates only a hypothetical justification for the observed facts. This means that the generated statements certainly imply the observed ones, but there may be also other explanations for the observed facts, while the generated statements do not hold.

As an example consider the following scheme

$$\frac{\text{rain} \Rightarrow \text{wet\_grass}, \text{wet\_grass}}{\text{rain}}.$$

Obviously, rain is a good reason for the grass being wet, but other explanations are possible; one of them can be that the sprinkler has just been used, another one can be the morning dew, etc. Abduction is a method used in diagnoses generation, both in technical systems and in the case of human diagnosis. Finally, abduction has something in common with inverted causal reasoning and guessing[8]. Abductive inference is applied in the so-called *backward-chaining* rule-based systems.

The basic abductive inference paradigm can be formalized as follows. Let $A_1, A_2, \ldots, A_n$ denote some set of assumptions (input knowledge), R be a set of rules, $B_1, B_2, \ldots, B_m$ be some observed or deduced knowledge about the considered system (output knowledge) and let Σ provide some additional observations (external knowledge). The basic formulation of abduction inference rule can be stated as follows

$$\frac{B_1, B_2, \ldots, B_m, R}{A_1, A_2, \ldots, A_n}$$

which is interpreted as: given the observation (output) $B_1, B_2, \ldots, B_m$ and the rules of R, find a possible set of assumptions $A_1, A_2, \ldots, A_n$, such that

$$A_1, A_2, \ldots, A_n \cup R \models B_1, B_2, \ldots, B_m.$$

Furthermore, the generated explanations $A_1, A_2, \ldots, A_n$ should be consistent with the auxiliary observations (if any) , i.e. $A_1, A_2, \ldots, A_n \cup \Sigma \cup R \not\models \bot$, i.e. consistency is preserved.

Note that abduction as defined above neither produces valid results ($A_1, A_2, \ldots, A_n$ is by no means the logical consequence of $B_1, B_2, \ldots, B_m, R$), nor provides unique result; in fact, there may be several (or none) possible results, satisfying the above conditions. Still, it seems to be widely applicable, in tasks such as diagnostics.

The domain knowledge R is expressed with a set of production rules, mostly of the form $q_1 \wedge q_2 \wedge \ldots q_k \longrightarrow h$. In such a case abductive inference is performed by matching the knowledge represented with $B_1, B_2, \ldots, B_m$ against specific rule conclusion and applying some backward chaining procedure until desired results are generated. Such a mechanism is applied e.g. in PROLOG [19, 24, 105] or most of backward-chaining rule-based systems. More complex systems may apply various strategies, both heuristic and based on optimality criteria, as well as auxiliary tools including testing and backtracking, consistency verification or meta-knowledge application.

[8] In fact, abduction was the favorite method used by Sherlock Holmes; contrary to what was written in the famous books by Sir Arthur Conan-Doyle, he applied abduction, and not deduction, to generate most of his so brilliant hypotheses so well explaining the unsolved mysteries.

The specific form of abduction makes it useful for tasks requiring hypotheses generation. Some most typical application examples include hypotheses generation, diagnostic inference, backward-chaining production systems, etc.

Various aspects and applications of abduction are discussed in [32].

1.9.2 Induction

Induction constitutes another inference paradigm, substantially different from deduction and abduction. To explain the basic idea of inductive inference assume that we observe several regularities representing facts, a relationship, dependencies or rules describing some events or items of the observed world. The task of inductive inference is to find a general theory covering (and implying) the observed individual or particular cases. The observed cases may be structured into positive and negative ones.

The generic induction task may be formalized as follows. Let *Th* denote a known domain theory and let R be a certain hypothesis to be generated. Further, let $C_1, C_2, \ldots, C_k$ denote observed specific cases. A general scheme of induction rules is as follows

$$\frac{C_1, C_2, \ldots, C_k;\ Th}{R},$$

where R is a general hypothesis rule, such that

$$Th \cup R \models C_1, C_2, \ldots, C_k\ ,$$

and perhaps many other new specific cases $C \notin \{C_1, C_2, \ldots, C_k\}$.

A more specific formulation of induction, especially in the context of rule-based systems, may be also as follows. Consider a set of examples in the form of simple inference or classification rules, $\alpha_1 \models h, \alpha_2 \models h, \ldots, \alpha_k \models h$. Each of these examples corresponds to a single specific case C_i, as described above. The rules provide information about decision, control action, classification etc. h which is common for a number of cases specified by the precondition α_i. It is assumed that they are detailed examples which can be covered by a single general rule. The induction task then lies in generating a rule constituting a rule-base R according to the following scheme

$$\frac{\alpha_1 \models h, \alpha_2 \models h, \ldots, \alpha_k \models h,\ Th}{R}.$$

The generated rule R should explain all the specific observations in the context of the assumed theory *Th*, i.e.

$$Th \cup R \models (\alpha_1 \models h, \alpha_2 \models h, \ldots, \alpha_k \models h)\ ,$$

for any $i = 1, 2, \ldots, k$. Simultaneously, R should be more concise (more abstract) than the detailed input rules.

In order to provide some intuitions let us consider a simple example. Let there be three observations, i.e.:

1) sparrow $\models$ flies,
2) pigeon $\models$ flies,
3) raven $\models$ flies.

and some theory $Th = \{\text{sparrow} \Rightarrow \text{bird}, \text{pigeon} \Rightarrow \text{bird}, \text{raven} \Rightarrow \text{bird}, \text{eagle} \Rightarrow \text{bird}, \ldots, \text{ostrich} \Rightarrow \text{bird}\}$. The induced general rule may be of the form

$$\text{bird} \Rightarrow \text{flies}.$$

Obviously, the rule explains and generalizes the above observations; it allows to predict some property of an *eagle* and perhaps many other birds, as well. However, induction as such is not a valid inference rule; applied to an *ostrich* it does not work. This is due to *over-generalization* which is an immanent feature of induction.

1.9.3 Deduction, Abduction and Induction — Mutual Relationship

To provide some intuitions about mutual relationship between deduction, abduction, and induction it may be convenient to consider classical black-box scheme for inference: imagine a black-box model containing rules R provided with some inputs A_i and producing outputs B_i. The task of deduction then is to generate (deduce) B_i being given A_i. The task of abduction constitutes an inverse problem: being given B_i find a possible explanation A_i, such that it implies B_i with respect to rules R. The task of induction is more general: being given specific inputs A_i and corresponding outputs B_i (forming the cases C_i) find a general theory usually in a form of a set of rules R describing all the observed cases; the theory provided with A_i should produce B_i and it should cover all the cases used for induction of R. It should also cover some new cases, since induction is the basic paradigm for learning systems [18]. Accessible background knowledge (if used) is denoted with Th.

In a number of specific applications induction is understood as generation of rules from a set of examples. This is, however, not the only induction possibility. Some other cases may concern generalization operations such as turning constants into variables, dropping conjunctive conditions or assigning disjunctive condition (see [53]), domain specific operations such as admitting interval instead of boundary values or turning an element or subfile into a larger file, or applying other generalization schemes, such as bd-resolution, described further in this chapter.

1.10 Generic Tasks of Propositional Logic

Propositional logic calculus is widely applied for modeling logical inference and specifying knowledge in various domains. Although its expressive power is limited to propositional symbols (no individual variables and more complex structures, such as terms are admitted) it can be used to encode practical knowledge in mathematical, technical and natural sciences. It can also serve as a generic tool for designing more elaborated, domain-specific languages.

Some generic tasks include specification and proving of theorems, verification of completeness, finding specific equivalent forms, especially minimal ones, and checking satisfiability of formulae. Below a short presentation of problem statement for these tasks at the level of propositional calculus is outlined.

1.10.1 Theorem Proving

Theorem proving is the most common task performed with propositional calculus and perhaps any other logic. Let H denote a formula specifying a certain theorem, and let Δ be a set of initial knowledge (knowledge base). The task is to prove that

$$\Delta \models H .$$

Basic approaches to theorem proving follow from Theorems 1 and 2 presented in Sect. 1.7. In automated theorem proving methods based on resolution (see Sect. 1.8.2) and on dual resolution (see Sect. 1.8.3) can be used. There exist many books on automated theorem proving, e.g. [7, 16, 36, 37, 39] and many other.

1.10.2 Tautology or Completeness Verification

Verification whether a formula is tautology may be regarded as a specific case of theorem proving. The task can be formalized by proving that

$$\models H .$$

In the context of rule-based systems, H may denote a set of formulae specifying the preconditions of rules operating in certain context. Such a set of formulae should have at least one rule to serve *any* potential input. This property is called *completeness* of a rule-based system.

Since in the above case H is naturally expressed in DNF, the dual resolution method constitutes a convenient tool for completeness verification [53, 57].

1.10.3 Minimization of Propositional Formulae

Looking for minimal representation of a logical formula is an important engineering task. This is applied in synthesis of electronic digital circuits, in synthesis of rule-based systems, PLC control systems and query optimization.

The generic task can be formalized as follows. Let Ψ denote some well-formed formula. The task is to find a formula Ψ', such that

$$\Psi' \equiv \Psi$$

and being in certain sense minimal (a simplest possible).

The methods for minimization are based on reduction of the initial formula; for intuition, this is done by appropriate transformation of the formula with the use of laws preserving logical equivalence, especially ones allowing for gluing components of the formula and by elimination of unnecessary components. It will be shown that dual resolution is a convenient tool for formula minimization.

The formula to be minimized is often presented in DNF. Then, the reduction is performed by subsequent application of dual resolution and elimination of subsumed minterms. Finally, the procedure should arrive at a point where a disjunction of some simplest (further irreducible) minterms is obtained. Such minterms are called *primary implicants.* Formally, we have the following definition.

Definition 25. *A primary implicant of a formula Ψ is a minterm π such that:*

- $\pi \models \Psi$,
- *there does not exist any minterm π' simpler than π ($[\pi'] \subset [\pi]$) such that $\pi' \models \Psi$.*

In fact, from logical point of view finding the minimal form of Ψ may be accomplished by finding a minimal set of its primary implicants, such that when put into a DNF they form a formula Ψ' which is logically equivalent to Ψ. The practical aspects of solving this task with dual resolution will be discussed in Chap. 15.

2

Predicate Calculus

Predicate Calculus is the most popular logical system, useful to formulate knowledge expressed with natural language (or restricted natural language), in the field of mathematics and in many other domains including computer science. In computer science *Predicate Calculus* is used for defining formal specifications and properties of programs and their components, and as a basic knowledge representation language in *Logic Programming* and in *Artificial Intelligence*. Thanks to its expressive power — which is obtained mainly due to admission of individual variables, terms and quantifiers, the language allows for practical formalization of quite complex knowledge.

Predicate Calculus (or *First Order Predicate Calculus*) inherits most of the basic ideas incorporated in propositional calculus. However, its expressive power is incomparably wider. Unfortunately, logical inference and theorem proving in predicate calculus become much more difficult, as well.

The name *Predicate Calculus* (or *First Order Predicate Calculus*, *FOPC*, for short) comes from the fact that this kind of logic is based on the use of *predicates* as the means for expressing knowledge about facts in a certain world under consideration. A *predicate* means a property or relation among some objects which are the arguments of it. A basic statement in predicate calculus is of the form $p(t_1, t_2, \ldots, t_n)$ and it has the meaning that *relation* p holds for objects $t_1, t_2, \ldots, t_n$; both names of precise objects as well as variables denoting *some* or *all* objects can be used. Finally, the calculus is of *first-order* since only basic objects can occur as arguments of relations and especially undergo quantification.

2.1 Alphabet and Notation

The alphabet of first-order logic consists of symbols denoting objects, functions and relations, logical connectives, quantifiers, and auxiliary elements, like parentheses and comma. The elements of the alphabet are presented below in detail.

A. Ligęza: *Logical Foundations for Rule-Based Systems*, Studies in Computational Intelligence (SCI) **11**, 37–50 (2006)
www.springerlink.com

Let there be given the following, pairwise disjoint four sets of symbols:

- C — a set of constant symbols (or constants, for short),
- V — a set of variable symbols (or variables, for short),
- F — a set of function (term) symbols,
- P — a set of relation (predicate) symbols.

All the sets are assumed to be countable (or finite, at least in specific applications).

Constants denote specific objects, items, elements, values, phenomena, etc. Variables are used to denote the same elements in case the precise name of an element is currently not known, unimportant, or a class of elements is to be represented.

2.1.1 The Role of Variables

The role of variables in first-order calculus is three-fold. It is worth examining the role in some details here, since it will influence design and properties of various classes of rule-based systems.

In short, variables place the role of:

- unknown but specific objects,
- place-holders,
- coreference constraints and data carriers.

First of all, variables may be used to denote *unknown but specific objects*; some variable $X \in V$ may denote an object the properties of which are specified without specifying the object by its name. This means that a class of objects can be defined in an implicit way, or one may refer to a group of objects using quantified variables.

Second, any functional and predicate symbol have assigned a constant number of arguments they operate on; this is called the *arity* of a symbol. Hence, the number of arguments cannot change — no argument can be missing. This means that if some of the arguments are unknown, variables must be used in place of specific names.

Last but not least, variables play the role of *coreference constraints* and data carriers. Two or more occurrences of the same variable in an expression denote the same object; if any replacement of an occurrence of some variable takes place, all the occurrences of this variable must be replaced with the same symbol or value. In this way data may be passed from rule input to output of the rule — a variable occurring in preconditions and conclusion of a rule will carry its value over the rule after being unified with some values during matching of preconditions against current state formula.

In the presented notation *variables* are denoted with single characters or strings, always beginning with an upper case letter or underscore, *constants* are denoted with any other strings of characters and special symbols or single letters (most typically a sequence of lower-case letters and possibly underscore characters; this convention is used in order to improve readability).

2.1.2 Function and Predicate Symbols

Function symbols denote, in general, mappings; however, in logic they are mostly applied to form record-like structures for representing more complex objects. In this case, one can assume that a function maps its arguments into the resulting structured object.

Predicate symbols are used to specify relations holding for certain objects.

It is assumed that for any functional symbol $f \in F$ and any predicate symbol $p \in P$ there is a unique function defining its number of arguments (arity) of the form $a\colon F \longrightarrow \{1, 2, 3, \ldots\}$ and $a\colon P \longrightarrow \{0, 1, 2, 3, \ldots\}$[1]. If for a certain function symbol f (predicate symbol p) the number of arguments is n, f (p) is called an n-place or n-ary function (predicate) symbol.

Functions and predicates are to be denoted with any arbitrary characters or strings; they are easily recognizable by their position in any expression. Further, proper names are frequently used so as to provide some intuitions and refer to some specific examples at hand. If necessary, indices can be occasionally used, so as to provide relatively precise definitions and theorems.

2.2 Terms in First-Order Logic

In order to denote any object — represented by a constant, a variable, or as a result of a mapping (a structured object), the notion of *term* is introduced. The set of terms *TER* is defined recursively in the following manner:

Definition 26. *The set of* terms *TER is one satisfying the following conditions:*

- *if c is a constant, $c \in C$, then $c \in$ TER;*
- *if X is a variable, $X \in V$, then $X \in$ TER;*
- *if f is an n-ary function symbol, $f \in F$, and $t_1, t_2, \ldots, t_n$ are terms, then $f(t_1, t_2, \ldots, t_n) \in$ TER;*
- *all the elements of TER are generated only by applying the above rules.*

The definition above imposes that only the expressions belonging to one of the above categories (i.e. constants, variables, and properly constructed structured objects) are terms. Furthermore, all of the expressions satisfying one of the above conditions are terms. Note that the definition is recursive, i.e. in order to check if a certain expression is a term one is to check if one of the above conditions holds; in case of the third possibility, the verifying procedure must be applied recursively down to all the elements $t_1, t_2, \ldots, t_n$, provided that f is an n-ary function symbol.

[1] By convention, functional symbols of no arguments are considered to be constants.

Assume that $a, b, c \in C$, $X, Y, Z \in V$, $f, g \in F$, and arity of f and g is 1 and 2, respectively. Then, all the following expressions are examples of terms:

- a, b, c;
- X, Y, Z;
- $f(a), f(b), f(c), f(X), f(Y), f(Z)$;
  $g(a, b), g(a, X), g(X, a), g(X, Y)$;
  $f(g(a, b)), g(X, f(X)), g(f(a), g(X, f(Z)))$.

Note that even for finite sets of constants, variables, and functions, it is possible to build an infinite set of terms (see examples in [16]). Obviously, if the set of functional symbols F is empty, the set of terms $TER = C \cup V$.

2.2.1 Applications of Terms

Terms can be used to represent various complex data structures, such as record-like objects, lists, trees, and many other. For intuition, let us show how general and flexible terms are in structure construction with a few examples.

Consider a book as an object having title, author, publisher, place and a year of publication. Further, let the author be a man having first name and surname. A book can be represented as a complex term of the form:

```
book (book_title,
      author(first_name,last_name),
      publisher_name,
      year_of_publication
     )
```

Note that many structures used in electronic documents, mathematics, formal languages and other systems are in fact terms. For example, in XML the example concerning the specification of a book can be represented as

```
<book>
      <book_title> Learning XML </book_title>
      <author>
            <first_name> Erik </first_name>
            <last_name>  Ray  </last_name>
      </author>
      <publisher_name> O'Reilly & Associates, Inc. </publisher_name>
      <year_of_publication> 2003 </year_of_publication>
</book>
```

where each field is declared in an explicit way by its name, beginning and end with a pair `<name> contents </name>` and the contents is either atomic value or another XML structure. Note that the internal structure of a tree is preserved.

Consider another example concerning specification of mathematical formulae. The following formula

$$\frac{\frac{x}{y}}{\sqrt{1+\frac{x}{y}}},$$

is in fact defined in LaTeX as

```
\frac{
      \frac{x}{y}
     }
     {
      \sqrt{1+\frac{x}{y}}
     }
```

where `\frac` is a two-argument symbol of division and `\sqrt` is a single argument symbol of $\sqrt{(.)}$.

Next, consider a list structure, e.g. `[red,green,blue,yellow]`. A list is constructed as an ordered pair of two elements: its *head*, being the single first element and its *tail* being the rest of the list (the definition is obviously recursive). A list as the one above can be represented as the following term

```
list(red,list(green,list(blue,list(yellow,nil))))
```

where `nil` is an arbitrary symbol denoting an empty list. Note that a list can be used to represent a set, a multi-set (or a bag — a set with repeated elements) and a sequence.

Finally, consider a binary tree, for example of depth 2; it can be represented by a term according to the following scheme

```
tree (
      tree (left_left,  left_right),
      tree (right_left, right_right)
     )
```

More complex trees can be represented with the use of lists, e.g. with a structure of the form

```
tree (root,list_of_subtrees)
```

Terms can be also used to specify graphs (e.g. as a list of nodes and another list of vertices), forests, relations, matrices, etc. In fact, expressive power of terms highly overcomes the immediate expectations following their definition. Some further examples will be presented in the part concerning PROLOG programming language.

2.3 Formulae

Formulae of first-order predicate logic are constructed in an analogous way to propositional logic; the main difference lies in the introduction of variables and quantifiers.

Predicate symbols are used to denote relations holding for certain objects. For this purpose a set *ATOM* of *atomic formulae*, is defined as follows.

Definition 27. *The set ATOM is defined as one satisfying the following conditions:*

- *if p is an n-ary predicate symbol, $p \in P$, and $t_1, t_2, \ldots, t_n$ are terms, then $p(t_1, t_2, \ldots, t_n) \in ATOM$;*
- *all the elements of ATOM are generated by applying the above rule.*

The elements of ATOM are called atomic formulae *or* atoms, *for short.*

Assume that p and q are predicate symbols of arity 1 and 2, respectively. Then, taking into account the above examples of terms, the following expressions are example atomic formulae:

- $p(a), p(b), q(a, a), q(a, c)$;
- $p(X), p(Y), q(X, X), q(X, Z)$;
- $p(f(a)), p(f(X)), q(f(g(a, b)), g(X, f(X))), q(g(f(a), g(X, f(Z))), a)$.

Atomic formulae are also referred to as *facts*, since they can be understood as single statements, i.e. if $p(t_1, t_2, \ldots, t_n)$ is an atomic formula, then it can be read as 'relation p holds for objects $t_1, t_2, \ldots, t_n$'.

More complex formulae can be created from the atomic ones with the use of the logical connectives. The set of well-formed formulae is defined in the following way.

Definition 28. *The set of formulae FOR is defined to be one satisfying the following conditions:*

- *$ATOM \subseteq FOR$;*
- *if Φ is a formula, $\Phi \in FOR$, then $\neg(\Phi) \in FOR$;*
- *if Φ and Ψ are formulae, $\Phi, \Psi \in FOR$, then $(\Phi \wedge \Psi)$, $(\Phi \vee \Psi)$, $(\Phi \Rightarrow \Psi)$, $(\Phi \Leftrightarrow \Psi) \in FOR$;*
- *if $\Phi \in FOR$, X denotes a variable, then $\forall X(\Phi) \in FOR$ and $\exists X(\Phi) \in FOR$;*
- *all the elements of FOR must be generated by applying the above rules.*

The elements of FOR are called formulae.

If we have in mind a specific set of constant, variable, function or predicate symbols occurring in some expression or a set of expressions E, we shall write $C(E)$, $V(E)$, $F(E)$, and $P(E)$, respectively.

An occurrence of a variable in an expression (formula) can be *free* or *bound.* The occurrence of variable X is *bound* if it is within the scope[2] of some quantifier $\forall X$ or $\exists X$. An occurrence of a variable which is not bound — i.e. one lying outside the scope of any quantifier referring to this variable — is *free.*

A variable is bound in a formula if at least one of its occurrences is bound in this formula. A variable is free in a formula if at least one of its occurrences is free in this formula. A precise definition of free variables is given below.

[2] The scope of quantifier $\forall$ ($\exists$) in $\forall X(\Phi)$ ($\exists X(\Phi)$) is the formula Φ.

Definition 29. *Let t be a term,* $t \in TER$*, and let q be an atomic formula,* $q \in ATOM$*. Let* Φ *and* Ψ *be some formulae,* $\Phi, \Psi \in FOR$*. Further, let* $FV(\mathrm{E})$ *denote the set of free variables in expression* E *(here: formula or term). The set* $FV(\mathrm{E})$ *is defined as follows:*

- *if* $t \in V$ *then* $FV(t) = \{t\}$;
- *if* $t \in C$ *then* $FV(t) = \emptyset$;
- *if* $t = f(t_1, t_2, \ldots, t_n) \in TER$ *then* $FV(t) = FV(t_1) \cup FV(t_2) \cup \ldots \cup FV(t_n)$;
- *if* $q = p(t_1, t_2, \ldots, t_n) \in ATOM$ *then* $FV(q) = FV(t_1) \cup FV(t_2) \cup \ldots \cup FV(t_n)$;
- $FV(\neg\Phi) = FV(\Phi)$;
- $V(\Phi \diamond \Psi) = FV(\Phi) \cup FV(\Psi)$ *for any* $\diamond \in \{\wedge, \vee, \Rightarrow, \Leftrightarrow\}$;
- $FV(\nabla X(\Phi)) = FV(\Phi) \setminus \{X\}$ *for* $\nabla \in \{\forall, \exists\}$.

Note that, according to the above definition, a variable can be both bound and free in a certain formula. Such a case takes place if some occurrences are bound and some of them are free. In such cases, in order to avoid ambiguities, we shall assume that a consistent renaming of the bound occurrences is applied, so that the free occurrences of the variable are the only ones remaining. Let for example $\Psi = p(X) \vee (\forall X\, q(X))$. In Ψ the first occurrence of X is free, while the second one is bound. Thus X is both free and bound in Ψ. After renaming the bound occurrence of X with Y we have $\Psi = p(X) \vee \forall Y q(Y)$, and thus X is only free and Y is only bound in Ψ.

Of course, such a renaming procedure does not change the meaning of the formula. Note also that although according to Definition 28 one can build formulae like $\forall X \Psi$ or $\exists X \Psi$ where X does not occur in Ψ as a free variable, it is impractical. A reasonable assumption is that if a certain variable is quantified in a formula, it must be free within the scope of the quantifier [16].

Note that, roughly speaking, for any term or formula, the set of free variables of it can be easily obtained by deleting from the set of all variables occurring in this term or formula the ones which are bound by the use of quantifiers.

2.4 Special Forms of Formulae

In the preceding section a general definition of well-formed formulae in first-order predicate calculus was given. According to this definition variety of logical formulae can be generated. However, in practical applications, it seems reasonable to define and make use of some *restricted* and *uniform* forms of formulae, as in the case of propositional logic.

The aim of such an approach is twofold. First, it is easier to implement computer programs dealing with limited forms of knowledge representation and, obviously, the resulting implementations are far more efficient. Second, theoretical properties of such restricted representations can be analyzed to a higher degree. Of course, the specific forms should not be too restrictive, so as not to loose generality.

Below, the definitions of some most important restricted forms of formulae are restated for first-order logic. Two of them, namely *clauses* and *Horn clauses* are typical for the *resolution* method of theorem proving [16, 39, 114] and PROLOG programming language based on it [11, 47, 105, 121]. Another two, namely *simple formulae* and *normal formulae* are typical for dual resolution theorem proving and modeling states and situations in dynamic systems.

Let us recall the definition of literals first.

Definition 30. *Let p be an atomic formula, $p \in ATOM$. Then, both p and $\neg p$ are called* literals*; p is called a* positive literal*, while $\neg p$ is called a* negative literal.

Thus a literal is an atom or a negated atom. Literals are components of clauses and minterms (simple formulae).

Definition 31. *Let $q_1, q_2, \ldots, q_k$ be literals of FOPC. A formula*

$$\varphi = q_1 \vee q_2 \vee \ldots \vee q_k$$

is called a clause *in FOPC.*

Thus, a clause is a finite disjunction of literals.

Definition 32. *Let $p_1, p_2, \ldots, p_k$ be positive literals. A* Horn clause *in FOPC is any clause of the form*

$$\varphi = \neg p_1 \vee \neg p_2 \vee \ldots \neg p_k \vee q \ ,$$

where q is a literal (either positive or negative one).

In other words, a Horn clause is any clause with at most one positive literal. Note also, that taking into account the De Morgan's laws and the definition of implication, a Horn clause can be represented as an implication of the form

$$\varphi = p_1 \wedge p_2 \wedge \ldots \wedge p_k \Rightarrow q \ .$$

Thus, a Horn clause can be regarded as a *rule*, having preconditions defined by $p_1 \wedge p_2 \wedge \ldots \wedge p_k$ and conclusion q. In fact, an efficient reasoning mechanism with the use of such rules is implemented within PROLOG [19, 105, 121].

Definition 33. *Let $q_1, q_2, \ldots, q_k$ be literals. Any formula of the form*

$$\phi = q_1 \wedge q_2 \wedge \ldots \wedge q_k$$

will be called an FOPC minterm or an FOPC simple formula.

Thus, a simple formula is a finite conjunction of literals. From now on, simple formulae will be most frequently denoted with the letters ϕ and ψ. Simple formulae play an important role in representation of state of dynamic systems; they are the core items for knowledge representation.

As in the case of propositional calculus, the use and therefore occurrence of negation sign in formulae is to a certain degree a matter of specific formalization of knowledge and can be modified with regard to the current area of application and user's preferences. For simplifying the notation one can often avoid the use of the negation symbol ($\neg$) in an explicit way. Instead of writing $\neg$`high(X)` one can rather write `low(X)`, instead of $\neg$`switch(on)` one can put `switch(off)`, etc., i.e. the negation can be often expressed implicitly.

More generally, in place of $\neg p$ one can always put a new atom, say np, which is logically equivalent to the negation of p with regard to the assumed interpretation. However, such an approach may lead to certain problems concerning automated reasoning, if no auxiliary rules (defining all the interdependencies among facts) are defined. For example, no purely logical inference engine will be capable of stating that a formula like `switch(on)` $\wedge$ `switch(off)` is always false — in such a case an auxiliary reasoning rule like, for example, `switch(off)` $\Rightarrow \neg$`switch(on)` should be provided so as to assure the detection of inconsistency.

As in the case of propositional logic, both a clause and a simple formula can be regarded as a set of its literals. It may be convenient then to apply the set notation directly to simple formulae. Recall that the set of literals from which a formula $\phi = q_1 \wedge q_2 \wedge \ldots \wedge q_k$ is composed is denoted as $[\phi] = \{q_1, q_2, \ldots, q_k\}$.

The definitions of Disjunctive Normal Form (DNF) and Conjunctive Normal Form (CNF) are analogous to the case of propositional calculus.

Definition 34. *Let $\phi_1, \phi_2, \ldots, \phi_n$ be some minterms (simple formulae). A formula of the form*

$$\Phi = \phi_1 \vee \phi_2 \vee \ldots \vee \phi_n$$

will be called an FOPC formula in Disjunctive Normal Form.

Definition 35. *Let $\psi_1, \psi_2, \ldots, \psi_n$ be some clauses. A formula of the form*

$$\Psi = \psi_1 \wedge \psi_2 \wedge \ldots \wedge \psi_n$$

will be called an FOPC formula in Conjunctive Normal Form.

Normal formulae can be regarded as sets of simple formulae. Again, the set notation can be applied directly to normal formulae if necessary.

Any formula containing no quantifiers can be effectively transformed into equivalent normal form (for the idea of such a transformation into *conjunctive normal form* see, for example [16]; transformation into disjunctive normal form is analogous (dual)). From now on, normal formulae will be most frequently denoted with the letters Φ and Ψ. Normal formulae in the presented approach correspond to the sets of clauses in resolution theorem proving.

2.5 Semantics of First-Order Logic

Formulae of first-order predicate calculus can be assigned a truth-value by defining an *interpretation* of the symbols of alphabet. In order to define an interpretation, the following three elements must be defined:

- a nonempty set D called a *universe*; the elements of D are the objects of some universe of discourse, in other words — the *domain* of interpretation;
- a mapping I which assigns meaning to constant, function, and predicate symbols, by providing the current understanding of these symbols with regard to the universe of discourse;
- a mapping v for assigning values to free variables.

Since in general free variables may occur in formulae, in order to be able to proceed, such variables must be assigned a meaning; this is done by defining a *variable assignment.*

Definition 36. *A* variable assignment v *is any mapping* $v\colon V \to D$.

Let D be a nonempty (finite) set, and let v be some variable assignment defined as above. Further, let I be a mapping of constant, function and predicate symbols into D, functions defined on D and relations on elements of D, respectively. The formal definition of *interpretation* is as follows.

Definition 37. *An* interpretation *of expressions of first-order predicate calculus is any mapping I operating on terms and formulae, satisfying the following conditions:*

- *for any constant* $c \in C$, $I(c) \in D$;
- *for any free occurrence of variable* $X \in V$, $I(X) = v(X)$, *where* $v(X) \in D$;
- *for any function symbol* $f \in F$ *of arity* n, $I(f)$ *is a function of the type*

$$I(f)\colon D^n \to D\ ;$$

- *for any predicate symbol* $p \in P$ *of arity* n, $I(p)$ *is a relation such that*

$$I(p) \subseteq D^n\ ;$$

- *for any term* $t \in TER$, *such that* $t = f(t_1, t_2, \ldots, t_n)$,

$$I(t) = I(f)(I(t_1), I(t_2), \ldots, I(t_n))\ .$$

Given an interpretation I (and variable assignment v, if necessary) over a domain D one can evaluate the *truth-value* of first order logic formulae. In order to do that, the idea of *satisfaction* for formulae is defined. For intuition, a formula is satisfied if and only if after consistent mapping of its elements into the universe of discourse, a true statement is obtained. In other case the formula is not satisfied, i.e. it represents a false statement with regard to the assumed interpretation.

As in the case of propositional calculus, the notion of *satisfaction* will be denoted with the standard symbol $\models$. Thus, $\models_{I,v} \Phi$ is to be read as 'Φ is true (is satisfied) under interpretation I for variable assignment v'. Similarly, $\not\models_{I,v} \Phi$ is to be read as 'Φ is false (is not satisfied) under interpretation I and variable assignment v'.

For simplicity, if either the variable assignment is not necessary (no free variables occur since all the formulae are closed ones) or the variable assignment is known by default, the explicit specification v may be omitted. The formal definition stating when a formula is satisfied is given below.

Definition 38. *Let I be an interpretation over domain D and let v be a variable assignment. Moreover, let $p(t_1, t_2, \ldots, t_n)$ be an atomic formula and let Φ and Ψ denote two formulae. The following statements define the satisfaction of formulae in FOR:*

1. $\models_{I,v} p(t_1, t_2, \ldots, t_n)$ *iff (if and only if)* $(I(t_1), I(t_2), \ldots, I(t_n)) \in I(p)$ *(recall that $I(X) = v(X)$ for any free variable $X \in VAR$;*
2. $\models_{I,v} \neg\Phi$ *iff* $\not\models_{I,v} \Phi$;
3. $\models_{I,v} \Phi \wedge \Psi$ *iff both* $\models_{I,v} \Phi$ *and* $\models_{I,v} \Psi$;
4. $\models_{I,v} \Phi \vee \Psi$ *iff* $\models_{I,v} \Phi$ *or* $\models_{I,v} \Psi$;
5. $\models_{I,v} \Phi \Rightarrow \Psi$ *iff* $\not\models_{I,v} \Phi$ *or* $\models_{I,v} \Psi$;
6. $\models_{I,v} \Phi \Leftrightarrow \Psi$ *iff* $\models_{I,v} \Phi$ *and* $\models_{I,v} \Psi$, *or,* $\not\models_{I,v} \Phi$ *and* $\not\models_{I,v} \Psi$;
7. $\models_{I,v} \forall X \Phi$ *iff for any $d \in D$ and any variable assignment u such that $u(X) = d$ and $u(Y) = v(Y)$ for any $Y \neq X$, there is* $\models_{I,u} \Phi$;
8. $\models_{I,v} \exists X \Phi$ *iff there exists $d \in D$ such that for variable assignment u defined as $u(X) = d$ and $u(Y) = v(Y)$ for any $Y \neq X$, there is* $\models_{I,u} \Phi$.

For convenience, one can say that 'a formula is satisfied under a given interpretation I' or, equivalently, that 'a given interpretation I satisfies a formula'.

The following definitions are re-stated copies of appropriate definitions introduced for propositional logic adapted for first-order logic.

Definition 39. *A formula is* consistent *(*satisfiable*) if and only if there exists an interpretation I over some domain D (and variable assignment v if necessary) which satisfies this formula.*

For simplicity, we shall also say that a formula is satisfiable, if given I and D one can choose a variable assignment v, such that the formula is satisfied under I.

Definition 40. *A formula is* inconsistent *(*unsatisfiable*) if and only if there does not exist any interpretation I and universe D (and variable assignment v, if necessary) satisfying the formula.*

Definition 41. *A formula is* valid *(*tautology*) if and only if for any interpretation I over any domain D there exists a variable assignment v, such that the formula is satisfied under the interpretation.*

Note that the above definition is equivalent to assuming that the free variables in a formula are implicitly existentially quantified; hence, for example, any formula of the form $p(X) \vee \neg p(Y)$ will be regarded as a valid one. Although in further parts of this work we shall refer to variable assignments or substitutions rather than to quantification, this kind of an implicit assumption will be prevalent for most of this work. Recall that a tautology is denoted as $\top$ while an unsatisfiable formula by $\bot$.

Definition 42. *A formula H is a* logical consequence *of formulae $\Delta_1, \Delta_2, \ldots, \Delta_n$ if and only if for any interpretation I satisfying $\Delta_1 \wedge \Delta_2 \wedge \ldots \wedge \Delta_n$ for some variable assignment v there exists a variable assignment u, such that H is satisfied under interpretation I and variable assignment u.*

Note that in most textbooks on logic and automated theorem proving the above definition [16] refers in fact only to the so-called *closed formulae*, i.e. ones with no free variables. In case free variables occur in a formula, one is to specify the way of 'understanding' them. Here, contrary to the most common approach [39], we insist on 'understanding' free variables as ones 'existentially quantified' and so we follow Definition 41.

Definition 43. *An interpretation I satisfying formula Φ is said to be a* model *for Φ.*

2.5.1 Herbrand Interpretation

To end with the basic notions of semantics we shall recall the ideas concerning the interpretation within the set of symbols introduced by the language, *i.e.* the Herbrand interpretation.

Definition 44. *Let $H_0 = C(\Delta)$, i.e. H_0 contains all the constants occurring in some set of formulae Δ (if $C(\Delta) = \emptyset$ then one defines H_0 in such a way that it contains a single arbitrary symbol, say $H_0 = \{c\}$). Now, for $i = 0, 1, 2, \ldots$, let $H_{i+1} = H_i \cup \{f(t_1, t_2, \ldots, t_n) \colon f \in F(\Delta) \text{ and } t_1, t_2, \ldots, t_n \in H_i\}$ (where the arity of f is n). Then H_∞ is called the* Herbrand universe *of Δ.*

Obviously, the Herbrand universe for a certain set Δ of formulae consists of all the ground terms (ones with no variables in them) which can be constructed with the use of the function symbols and constants occurring in the formulae of Δ[3]. For simplicity, instead of H_∞ we shall also write $\mathbf{H}$.

Definition 45. *Let Δ be a set of formulae and let $\boldsymbol{H}$ be the Herbrand universe of Δ. A set $B_H = \{p(h_1, h_2, \ldots, h_n) \colon h_1, h_2, \ldots, h_n \in \boldsymbol{H}, p \in P(\Delta)\}$ (where the arity of p is n) is called the* Herbrand base *or the* atom set *of Δ.*

[3] The intuition behind the Herbrand Universe is as follows: instead of considering all possible universes, it is enough to restrict the analysis to such universes, that the elements of them can be named with the use of the symbols of the language.

Obviously, the Herbrand base for a given set Δ of formulae is the set of all the atomic formulae with no variables (ground atoms), which can be constructed from the predicate symbols occurring in Δ and ground terms belonging to the Herbrand universe of Δ.

Definition 46. *Let Δ be a set of formulae and let $\boldsymbol{H}$ be the Herbrand universe of Δ. Any interpretation I_H is called a* Herbrand interpretation *(**H**-interpretation) if the following conditions are satisfied:*

- *for any constant $c \in \boldsymbol{H}$, $I_H(c) = c$;*
- *for any n-ary functional symbol $f \in F(\Delta)$, and any $h_1, h_2, \ldots, h_n \in \boldsymbol{H}$,*

$$I_H(f) \colon (h_1, h_2, \ldots, h_n) \rightarrow f(h_1, h_2, \ldots, h_n)\,.$$

An **H**-interpretation is any interpretation mapping the ground terms into the same ground terms. The importance of the concept of **H**-interpretations follows from the fact that they 'cover' other interpretations in certain cases. For example, given a ground formula and any interpretation I of this formula, one can build a corresponding **H**-interpretation (see [16]) assigning the same truth value to all the literals of the formula and to the formula as a whole. For intuition, this follows from the fact that being given a set of constants and function symbols occurring in a formula, one can 'name' only as many objects as there are in the Herbrand universe.

Note that disregarding the variable assignment v, for any set of formulae Δ, a given **H**-interpretation I_H of Δ can be conveniently represented by explicit listing of the literals satisfied by the interpretation (evaluated to *true* under this interpretation). Thus, the **H**-interpretation can be given as a set $I_H = \{q_1, q_2, \ldots, q_n, \ldots\}$, where q_i is either an element of H_B or a negation of such an element; of course, if some $q \in I_H$, then $\neg q \notin I_H$, since an interpretation cannot simultaneously assign true and false to any formula.

Below we present a version of the Herbrand theorem [16]; in fact, the presented theorem is a slightly generalized dual version of the original.

Theorem 5 (Herbrand Theorem). *Let Ψ be a normal DNF formula and ψ' a ground simple formula. Then $\psi' \models \Psi$ if and only if there exists a ground normal formula Ψ' being a substitution instance of Ψ, such that $\psi' \models \Psi'$.*

Proof. Assume that $\psi' \models \Psi'$, where Ψ' is a ground substitution instance of Ψ. Obviously, $\psi' \models \Psi$ since while passing from Ψ' to Ψ one is to turn some constants into variables, and thus 'improve' the generality of the formula. Now we shall prove the other part of the theorem.

If ψ' is unsatisfiable, then any ground instance of Ψ satisfies the theorem. Assume that ψ' is satisfiable. Since ψ' is a ground formula, it is enough to consider only **H**-interpretations satisfying it (one cannot assign more objects to the terms of ψ' than one can find in the Herbrand universe of ψ'). Let I be any Herbrand interpretation satisfying ψ', i.e. such that $\models_I \psi'$. Interpretation

I takes as a universe $\mathbf{H}$ denoting the Herbrand universe of ψ' (or a larger set), I is any mapping satisfying the definition of interpretation and such that all the literals of ψ' are satisfied. Taking into account that $\psi' \models \Psi$, there exists a variable assignment u, such that for interpretation I with u there is $\models_{I,u} \Psi$. Now, let us replace the variables of Ψ with the elements of $\mathbf{H}$ to which the variables are mapped with u; in this way Ψ' being a ground formula is obtained. Obviously $\models_I \Psi'$. Since the choice of u is influenced only by the selection of $\mathbf{H}$ (defined by ψ') and the requirement that I should be chosen so as to satisfy all the literals of ψ' (invariant for single ψ'), once obtained, the formula Ψ' will be satisfied under any interpretation satisfying ψ'. Hence $\psi' \models \Psi'$. □

3

Attribute Logic

Attribute Logic (AL, for short) is one based on the use of attributes for denoting some properties of objects and a system under consideration. In order to define characteristics of the system one selects some specific set of attributes and assigns them some values. This way of describing an object and system properties is both simple and intuitive.

Using logic based on attributes is one of the most popular approaches to define facts about a certain system. This kind of logic is omnipresent in various applications and domain-specific formalisms, such as attributive decision tables, decision trees, attributive rule-based systems and even relational databases. In fact, a table in relational database can be considered to represent knowledge of an attributive logic formula.

In computer sciences attribute based languages are used for defining formal specifications and properties of programs and their components. In relational databases formulae of attributive logic are used to define selection criteria for information retrieval and for the so-called θ-join operation for joining tables [22]. Thanks to its expressive power — greater than in the case of propositional calculus — attributive languages allow for practical formalization of quite complex knowledge.

Attributive logic inherits most of the basic ideas incorporated in propositional calculus. However, its expressive power is grater, depending on the admitted values of attributes, relational symbols and variables. Unfortunately, both syntax and semantics, as well as theorem proving in attributive calculi become a bit more difficult, as well.

The name *Attributive Logic* (or *Attribute-Based Logic*) comes from the fact that this kind of logic is based on the use of *attributes* that have some values assigned as the means for expressing knowledge about facts in a world under consideration. An *attribute* means a property or a characteristic feature taking a certain defined or unknown value at a certain instant of time. For intuition, a basic statement in attributive calculus is of the form $\langle \mathit{attribute} \rangle (\langle \mathit{object} \rangle) = \langle \mathit{value} \rangle$ or $A(o) = v$ and it has the meaning that attribute A for object o takes value v. For example, $\mathit{Color}(\mathit{car}) = \mathit{red}$ means that the color of a specific car

A. Ligęza: *Logical Foundations for Rule-Based Systems*, Studies in Computational Intelligence (SCI) 11, 51–63 (2006)
www.springerlink.com

is red. In such a way, values of selected properties are assigned to an object of interest.

There are various possibilities of defining attribute-based languages; depending on the required expressive power, the following possibilities will be considered:

- AAL — *Atomic Attributive Logic*, i.e. attributive logic with atomic values of attributes only;
- SAL — *Set Attributive Logic*, i.e. attributive logic with set values of attributes;
- VAAL — *Variable Atomic Attributive Logic*, i.e. attributive logic with atomic values of attributes incorporating variables;
- VSAL — *Variable Set Attributive Logic*, i.e attributive logic with set values of attributes incorporating variables.

Moreover, attribute-based languages may allow the use of other than equality relational symbols, i.e. $>$, $\geq$, etc.; however, this issue will not be described here. Instead, we shall show a way of dealing with such relations through replacing the constraints based on different symbols with equality and using set values.

3.1 Alphabet and Notation

The alphabet of attributive logic consists of symbols denoting objects, attributes, constant values, variables, logical connectives, quantifiers, and auxiliary elements, like parentheses and comma. Below the elements of the alphabet are presented in some details.

Let there be given the following, pairwise disjoint sets of symbols:

- O — a set of object name symbols,
- A — a set of attribute names,
- D — a set of attribute values names (the *domains*),
- V — a set of variable symbols (or variables, for short).

All the sets are assumed to be countable (or finite, at least in specific applications).

Constant values belonging to domain D denote values of specific attributes for given objects. Variables are used to denote the same elements in case the precise name of an element is currently unknown, unimportant, or a class of elements is to be represented.

It is further frequently assumed that the domain is divided into several sets (disjoint or not), such that any of the sets defines the domain of some attribute. More precisely, if the set of attributes is finite and given as

$$A = \{A_1, A_2, \ldots, A_n\} ,$$

then also

$$D = D_1 \cup D_2 \cup \ldots \cup D_n ,$$

where D_i is the domain of attribute A_i, $i = 1, 2, \ldots, n$.

For further use the following basic definition of attribute is assumed.

Definition 47. *An attribute A_i is a function (or partial function) of the form*

$$A_i \colon O \to D_i .$$

Such a definition of attribute requires that an attribute can take a single value for an object at a time. An attribute can be a *partial function*, since not all the attributes must be defined for all the objects. Such a definition is used in the attributive languages operating on single values, i.e. AAL and VAAL (see Sect. 3.2).

For more powerful languages, a more general definition of an attribute (generalized attribute) is necessary; such an attribute can take *several values* at a time.

Definition 48. *A generalized attribute A_i is a function (or partial function) of the form*

$$A_i \colon O \to 2^{D_i} .$$

The generalized attributes are used in languages such as SAL and VSAL (see Sect. 3.2). If it does not introduce ambiguity, the qualifier *generalized* will frequently be omitted.

3.1.1 The Role of Variables

The role of variables in attributive logic calculus is two-fold. It is worth examining the role in detail here, since it will influence design and properties of various classes of rule-based systems.

In short, variables play the role of:

- undefined, unknown but specific values,
- coreference constraints and data carriers.

First of all, variables may be used to denote *unknown but specific values*; a variable $X \in V$ may denote a certain value, perhaps currently unknown. It may be further specified by instantiation — replacing the variable with a specific value.

Secondly, variables play the role of coreference constraints and data carriers. Two or more occurrences of the same variable in an expression denote the same object; if any replacement of an occurrence of some variables takes place, all the occurrences of this variable must be replaced with the same symbol or value. In this way data may be passed from rule input to output of the rule — a variable occurring in preconditions and conclusion of a rule will carry its value over the rule after being unified with some values during matching of precondition. This is also the way of defining rules with conclusions depending on the values of some attributes in the precondition part in a certain functional way (as, e.g. in the case of defining tax on the base of the income).

In the presented notation *variables* are denoted with single characters or strings, always beginning with an upper case letter or underscore, *constant atomic values* are denoted with any other strings and characters (usually as $d \in D$) and set values in a similar way (e.g. $t \subseteq D$).

3.2 Atomic Formulae

Formulae of attributive logic are constructed in a way that is analogous to propositional logic; the main difference lies in the introduction of variables.

Attribute symbols are used to denote properties of certain objects. To denote the fact that a certain attribute takes a certain value one constructs an atomic formula. The set $ATOM$ being the set of *atomic formulae*, is defined in the following way.

Definition 49 (AAL). *Let $o \in O$ be a certain object, $A_i \in A$ be an attribute and let $d \in D_i$ be a certain atomic value of the domain of A_i. Any expression of the form*

$$A_i(o) = d$$

is an atomic formula of AAL.

Definition 50 (SAL). *Let $o \in O$ be some object, $A_i \in A$ be an attribute and let $t \subseteq D_i$ be a certain subset of the domain of A_i. Any expression of the form:*

$$A_i(o) = t$$

and

$$A_i(o) \in t$$

are atomic formulae of SAL.

Note that the definition of atomic formulae in SAL (50) covers the one in AAL (49); in fact, any atomic value can be considered as a single-element set. The vice versa is obviously not true. For simplicity, if the object is known, the formulae are simplified to $A_i = d$, $A_i = t$ or $A_i \in t$, respectively. Without object specification, such simplified formulae are called *selectors* since they can be used for selecting a set of objects satisfying the specific condition. They are often denoted as $[A_i = d]$ and $[A_i = t]$ and $[A_i \in t]$.

The cases of VAAL and VSAL, i.e. when variables are also allowed require some extended definitions.

Definition 51 (VAAL). *Any atomic formula of AAL is also and atomic formula of VAAL. Further, if $o \in O$ is a certain object, $A_i \in A$ is an attribute and $X \in V$ is a certain variable, then any expression of the form:*

$$A_i(o) = X$$

is an atomic formula of VAAL.

Definition 52 (VSAL). *Any atomic formula of SAL is also and atomic formula of VSAL. Further, if $o \in O$ is a certain object, $A_i \in A$ is an attribute and $X \in V$ is a certain variable, then any expression of the form:*

$$A_i(o) = X$$

and

$$A_i(o) \in X$$

is an atomic formula of VSAL.

It is also assumed that all atomic formulae must be defined as in one of the above definitions, i.e. Def. 49, 50, 51 or 52; in other words no other expression is an atomic formula.

Atomic formulae are also referred to as *facts*, since they can be understood as single statements, i.e. if $A_i(o) = d$ is an atomic formula, then it can be read 'a fact that the value of attribute A_i for object o is d holds'.

3.3 Formulae in Attribute Logic

More complex formulae can be generated from the atomic ones with the use of the logical connectives and quantifiers. The set of well-formed formulae *FOR* is defined in the following way.

Definition 53. *The set of formulae FOR is defined to be one satisfying the following conditions:*

- $ATOM \subseteq FOR$;
- *if Φ is a formula, $\Phi \in FOR$, then $\neg(\Phi) \in FOR$;*
- *if Φ and Ψ are formulae, $\Phi, \Psi \in FOR$, then* $(\Phi \wedge \Psi)$, $(\Phi \vee \Psi)$, $(\Phi \Rightarrow \Psi)$, $(\Phi \Leftrightarrow \Psi) \in FOR$;
- *if $\Phi \in FOR$, X denotes a variable, then $\forall X(\Phi) \in FOR$ and $\exists X(\Phi) \in FOR$;*
- *all the elements of FOR must be generated by applying the above rules.*

The elements of FOR are called formulae.

If we have in mind a specific set of constant values, variables, or object names occurring in an expression or a set of expressions E, we shall write $D(E)$, $V(E)$, and $O(E)$, respectively.

An occurrence of a variable in an expression (formula) can be *free* or *bound.* The occurrence of variable X is *bound* if it is within the scope of some quantifier $\forall X$ or $\exists X$. An occurrence of a variable which is not bound — i.e. one lying outside the scope of any quantifier referring to this variable — is *free.*

A variable is bound in a formula if at least one of its occurrences is bound in this formula. A variable is free in a formula if at least one of its occurrences is free in this formula. A precise definition of free variables is given below.

Definition 54. *Let q be an atomic formula, $q \in ATOM$. Let Φ and Ψ be some formulae, $\Phi, \Psi \in FOR$. Further, let $FV(\mathrm{E})$ denote the set of free variables in expression* E *(here: formula or any value or variable). The set $FV(\mathrm{E})$ is defined as follows:*

- *if $X \in V$ then $FV(X) = \{X\}$,*
- *if $d \in D$ then $FV(d) = \emptyset$,*
- *if $t \subseteq D$ then $FV(t) = \emptyset$,*
- *if q is defined as $A_i(o) = e$ or $A_i(o) \in e$, $q \in ATOM$ then $FV(q) = FV(e)$,*
- *$FV(\neg\Phi) = FV(\Phi)$,*
- *$FV(\Phi \diamond \Psi) = FV(\Phi) \cup FV(\Psi)$ for any $\diamond \in \{\wedge, \vee, \Rightarrow, \Leftrightarrow\}$,*
- *$FV(\nabla X(\Phi)) = FV(\Phi) \setminus \{X\}$ for $\nabla \in \{\forall, \exists\}$.*

The definition above may be considered to be to complex; it is put in such a way in order to keep a unique way of defining the set of free variables in case of AL and predicate logic.

Note that, according to the above definition, a variable can be both bound and free in certain formula. Such a case takes place if some occurrences are bound and some of them are free. In such cases, in order to avoid ambiguities, we shall assume that a consistent renaming of the bound occurrences is applied, so that the free occurrences of the variable are the only ones remaining. Let for example $\Psi = ((A(o) = X) \vee (\forall X B(o) = X))$. In Ψ the first occurrence of X is free, while the second one is bound. Thus X is both free and bound in Ψ. After renaming the bound occurrence of X with Y we have $\Psi = ((A(o) = X) \vee \forall Y B(o) = Y)$, and thus X is only free and Y is only bound in Ψ.

Of course, such a renaming procedure does not change the meaning of the formula. Note also that although according to Definition 54 one can build formulae like $\forall X \Psi$ or $\exists X \Psi$ where X does not occur in Ψ as a free variable (especially in the case of variable-free languages, i.e. AAL and SAL), it seems to be impractical. A reasonable assumption is that if a certain variable is quantified in a formula, it must be free within the scope of the quantifier [16].

Note that, roughly speaking, for any term or formula, the set of free variables of it can be easily obtained by deleting from the set of all variables occurring in this term or formula the ones which are bound by use of quantifiers.

3.4 Semantics of Attribute Logic

Formulae of attributive logic can be assigned truth-value by defining an *interpretation* of the symbols of alphabet. In order to define an interpretation, the following four elements must be defined:

- a nonempty set U called a *universe*; the elements of U are the objects of a certain universe of discourse, in other words — the *domain* of interpretation;
- a nonempty set U_o of objects of interest;
- a mapping I which assigns meaning to atomic constants, object and attribute symbols, by providing the current understanding of these symbols with regard to the universe of discourse;
- a mapping v for assigning values to free variables.

Since in general free variables may occur in formulae, in order to be able to proceed, such variables must be assigned a meaning; this is done by defining a *variable assignment.*

Definition 55. *A* variable assignment v *is any mapping* $v\colon V \to U \cup 2^U$.

Let U be a nonempty (finite) set, and let v be some variable assignment defined as above. Further, let I be a mapping of atomic constants, object and attribute symbols into U, objects of U_o and functions on elements of U_o with values in U, respectively. The formal definition of *interpretation* is as follows.

Definition 56. *An* interpretation *of expressions of AL formula is any mapping I extended over formulae, satisfying the following conditions:*

- *for any constant symbol* $d \in D$, $I(d) \in U$;
- *for any object symbol* $o \in O$, $I(o) \in U_o$;
- *for any free variable* $X \in V$, $I(X) = v(X)$, *where* $v(X) \in U \cup 2^U$;
- *for any set symbol* $t \subseteq D$, $I(t) \subseteq U$;
- *for any attribute name* $A_i \in A$, $I(A_i)$ *is a (partial) function of the type*

$$I(A_i)\colon U_o \to U \cup 2^U .$$

For the sake of consistency, it is assumed that if $d \in t$, then also $I(d) \in I(t)$ and vice versa.

Given an interpretation I (and variable assignment v, if necessary) over a domain U one can evaluate the *truth-value* of AL formulae. In order to do that, the idea of *satisfaction* for formulae is defined. For intuition, a formula is satisfied if and only if after consistent mapping of its elements into the universe of discourse, a true statement is obtained. In other case the formula is not satisfied, i.e. it represents a false statement with regard to the assumed interpretation.

The notion of *satisfaction* will be denoted with the standard symbol $\models$. Thus, $\models_{I,v} \Phi$ is to be read as 'Φ is true (is satisfied) under interpretation I for variable assignment v'. Similarly, $\not\models_{I,v} \Phi$ is to be read as 'Φ is false (is not satisfied) under interpretation I and variable assignment v'.

For simplicity, if either the variable assignment is not necessary (no free variables occur since all the formulae are closed ones) or the variable assignment is known by default, the explicit specification of v may be omitted. The formal definition stating when a formula is satisfied is given below.

Definition 57. *Let I be an interpretation over domain U and let v be a variable assignment. Moreover, let $A_i(o) = e$ be an atomic formula and let Φ and Ψ denote two formulae. The following statements define the satisfaction of formulae in FOR:*

1. $\models_{I,v} A_i(o) = e$ *iff (if and only if)* $I(A_i)(I(o)) = I(e)$ *(recall that* $I(X) = v(X)$ *for any free variable* $X \in V$*;*
2. $\models_{I,v} A_i(o) \in e$ *iff (if and only if)* $I(A_i)(I(o)) \in I(e)$ *(recall that* $I(X) = v(X)$ *for any free variable* $X \in V$*;*
3. $\models_{I,v} \neg\Phi$ *iff* $\not\models_{I,v} \Phi$*;*
4. $\models_{I,v} \Phi \wedge \Psi$ *iff both* $\models_{I,v} \Phi$ *and* $\models_{I,v} \Psi$*;*
5. $\models_{I,v} \Phi \vee \Psi$ *iff* $\models_{I,v} \Phi$ *or* $\models_{I,v} \Psi$*;*
6. $\models_{I,v} \Phi \Rightarrow \Psi$ *iff* $\not\models_{I,v} \Phi$ *or* $\models_{I,v} \Psi$*;*
7. $\models_{I,v} \Phi \Leftrightarrow \Psi$ *iff* $\models_{I,v} \Phi$ *and* $\models_{I,v} \Psi$*, or,* $\not\models_{I,v} \Phi$ *and* $\not\models_{I,v} \Psi$*;*
8. $\models_{I,v} \forall X\Phi$ *iff for any* $u \in U \cup 2^U$ *and any variable assignment* w *such that* $w(X) = u$ *and* $w(Y) = v(Y)$ *for any* $Y \neq X$*, there is* $\models_{I,w} \Phi$*;*
9. $\models_{I,v} \exists X\Phi$ *iff there exists* $u \in U \cup 2^U$ *such that for variable assignment* w *defined as* $w(X) = u$ *and* $w(Y) = v(Y)$ *for any* $Y \neq X$*, there is* $\models_{I,w} \Phi$*.*

For convenience, one can say that 'a formula is satisfied under a given interpretation I' or, equivalently, that 'a given interpretation I satisfies a formula'.

The following definitions are re-stated versions of appropriate definitions introduced for propositional logic.

Definition 58. *A formula is* consistent *(*satisfiable*) if and only if there exists an interpretation I over some domain U (and variable assignment v if necessary) which satisfies this formula.*

For simplicity, we shall also say that a formula is satisfiable, if given I and U one can choose a variable assignment v, such that the formula is satisfied under I.

Definition 59. *A formula is* inconsistent *(*unsatisfiable*) if and only if there does not exist any interpretation I (and variable assignment v) satisfying the formula.*

Definition 60. *A formula is* valid *(*tautology*) if and only if for any interpretation I over any domain U there exists a variable assignment v, such that the formula is satisfied under the interpretation.*

Note that the above definition is equivalent to assuming that the free variables in a formula are implicitly existentially quantified; hence, for example, any formula of the form $A_i = Y \vee \neg(A_i = X)$ will be regarded as a valid one. Although in further parts of this work we shall refer to variable assignments or substitutions rather than to quantification, this kind of an implicit assumption will be prevalent for most of this work. Recall that a tautology is denoted as $\top$ while an unsatisfiable formula by $\bot$.

Definition 61. *A formula H is a* logical consequence *of formulae $\Delta_1, \Delta_2, \ldots, \Delta_n$ if and only if for any interpretation I satisfying $\Delta_1 \wedge \Delta_2 \wedge \ldots \wedge \Delta_n$ for a certain variable assignment v there exists a variable assignment w, such that H is satisfied under interpretation I and variable assignment w.*

Note that in most textbooks on logic and automated theorem proving the above definition [16] refers in fact only to the so-called *closed formulae*, i.e. ones with no free variables. In case free variables occur in a formula, one is to specify the way of 'understanding' them. Here, contrary to the most common approach [39], we insist on 'understanding' free variables as ones 'existentially quantified' and so we follow Definition 60.

Definition 62. *An interpretation I satisfying formula Φ is said to be a* model *for Φ.*

3.5 Issues Specific to Attribute-Based Logic

There are several issues specific to attribute-based logic. Unfortunately, they are not covered by majority of textbooks on logic. Below, a selection of such issues is presented in detail.

3.5.1 Internal Conjunction

First, consider the case of discrete, finite domains. Using the language of AAL attributes can take only atomic values. After moving to a more expressive language of SAL, attributes can take set values. This means that such an attribute can take more than one value at a time for a given object. Such attributes are specific[1] but useful in numerous practical applications. Especially when the set of values assigned to an object contains numerous values the notation $A(o) = t$ can be considered as a useful shorthand.

Consider a typical example of specifying foreign languages known by people — most of the people do not know any foreign language, some of

[1] Sometimes such attributes are called multiple valued attributes.

them know one or two, but there are people who know ten or more languages. This problem encountered in classical RDBS with atomic values leads to the so-called *fourth normal form* [23]. An atom specifying that an attribute takes set value can be represented in a (logically equivalent) form incorporating atomic values only. The set representation of several atomic values is also referred to as *internal conjunction*. Thus instead of writing $\mathit{kfl}(\mathit{doe}) = \mathit{english} \wedge \mathit{kfl}(\mathit{doe}) = \mathit{french} \wedge \mathit{kfl}(\mathit{doe}) = \mathit{spanish}$ one can simply write $\mathit{kfl}(\mathit{doe}) = \{\mathit{english}, \mathit{french}, \mathit{spanish}\}$, where $\mathit{kfl}(X)$ stays for 'knows-foreign-languages(X)'.

In fact, in the atomic representation in AAL, where attributes can take atomic values, a conjunctive formula as below can be transformed into a single atom of SAL according to the following principle

$$[(A_i = d_1) \wedge (A_i = d_2) \wedge \ldots \wedge (A_i = d_j)] \equiv [A_i = t] \ ,$$

where $t = \{d_1, d_2, \ldots, d_j\}$ is a subset of D[2]. An analogous extension applies to interval representation. For example, $A_i = [a, b]$ means that all the values belonging to the interval $[a, b]$ are covered (both in case of discrete and continuous domain of the attribute). In case of infinite set t, the atom $A_i(o) = t$ does not have finite internal conjunction form replacing it.

3.5.2 Internal Disjunction

A similar problem to the above occurs when one has to specify a long disjunctive formula specifying *possible* atomic values for the same object and attribute. Let us introduce a new relational symbol $\in$ having the obvious meaning. In fact, in the atomic representation in AAL, where attributes can take atomic values, a disjunctive formula as below can be transformed into a single atom of SAL according to the following principle

$$[(A_i = d_1) \vee (A_i = d_2) \vee \ldots \vee (A_i = d_j)] \equiv [A_i \in t] \ ,$$

where $t = \{d_1, d_2, \ldots, d_j\}$ is a subset of D[3]. An analogous extension applies to interval representation. For example, $A_i \in [a, b]$ means that all the values belonging to the interval $[a, b]$ are possible values for A_i (both in case of discrete and continuous domain of the attribute).

Note that the above transformation can be applied only in case of finite domains. In case of infinite domains, both countable and continuous, the set notation of SAL has no equivalent in the language of AAL. Hence, the expressive power of SAL is higher than the one of AAL.

[2] According to the rules of syntax defined above and for simplicity, the notation $A_i(o) = t$ is used throughout this book.

[3] According to the rules of syntax defined above and for simplicity, the notation $A_i(o) \in t$ is used instead for convenience.

3.5.3 Explicit and Implicit Negation

Although negation can be represented explicitly, it is often the case that the explicit use of it is avoided. In such a case we say that we use only *positive representation* and *positive atomic formulae.* A negated atomic formula can be transformed into a positive one according to the following rule

$$\neg(A_i = d) \equiv (A_i \in \overline{d}) ,$$

where $\overline{d}$ is the complement of d, i.e. $\{d\} \cup \overline{d} = D_i$, $\{d\} \cap \overline{d} = \emptyset$; in fact $\overline{d} = D \setminus \{d\}$. Unfortunately, for set values the transformation becomes more clumsy and requires referring to specific values of t.

In case of internal disjunction we have the following transformation

$$\neg(A_i \in t) \equiv (A_i \in \overline{t}) ,$$

where $\overline{t}$ is the complement of t, i.e. $t \cup \overline{t} = D_i$, $t \cap \overline{t} = \emptyset$; in fact $\overline{t} = D \setminus t$.

Let $\_$ denote any value of an appropriate domain[4]. Note that there is also

$$(A_i \in \_) \equiv (A_i \in D_i)$$

having the consequence that $\neg(A_i \in \_) \equiv (A_i \in \emptyset)$ (always false).

Specific problems may occur in case of set values being intervals. For example, the negation of belonging to a convex interval situated somewhere in a middle of the domain may lead to the sum of the left and right complement of it. Let $D_i = [0, 10]$. We have

$$\neg(A_i \in [3, 7]) \equiv A_i \in [0, 3) \cup (7, 10] ,$$

i.e. in case of intervals reduction of negation may lead to complex values, such as sum of convex intervals (sometimes called *non-convex intervals*).

For the sake of simplicity, in this book we do not allow for relation symbols different than equality ($=$) and 'belongs to' ($\in$), at least in the basic presentation of attribute-based logic. However, depending on the current needs, one can admit various notational possibilities extending the syntax (and semantics) of the AL. For example, if the set of values for attribute A_i is an ordered set, one can use typical algebraic symbols such as $<$, $>$, $\leq$, $\geq$. For instance, if $D_i = \{0, 1, 2, 3, 4, 5, 6, 7, 8, 9\}$, then $A_i \in \{0, 1, 2, 3\}$ can be denoted as $A_i \leq 3$, and $A_i \in \{3, 4, 5\}$ can be denoted as $3 \leq A_i \leq 5$. Further, $A_i = \{3, 4, 5\}$ can be denoted as $A_i = [3, 5]$, etc. Such use of relational symbols for specification of atomic formulae (or selectors) can be considered as popular shorthand and in fact it is very common in the domain literature.

---

[4] This is the convention used in PROLOG programming language.

3.6 Inference Rules Specific to Attributive Logic

Apart from the fact that all the logical equivalences and inference rules introduced for propositional calculus are also valid for attribute-based logic, there are some rules specific to logic based on attributes. These rules follow from the properties and issues observed in the former section and from the interpretation of AL formulae. Fortunately, all of them are simple and intuitive.

Firstly, consider two subset symbols $s, t \subseteq D$. Obviously, for any object symbol $o \in O$ and any attribute $A_i \in A$, if t is a subset of s the following rule holds

$$\frac{A_i(o) = s}{A_i(o) = t}\,. \tag{3.1}$$

Rule (3.1) will be referred to *downward consistency rule* or *subset consistency rule.* The proof of this rule follows immediately from the definition of interpretation and the assumption that if $d \in t$, then also $d \in s$. The meaning of the downward consistency rule is obvious — if an attribute takes values from a certain set, then certainly its values stay within any subset of that set.

By analogy, consider two subset symbols $s, t \subseteq D$. Obviously, for any object symbol $o \in O$ and any attribute $A_i \in A$, if s is a subset of t the following rule holds

$$\frac{A_i(o) \in s}{A_i(o) \in t}\,. \tag{3.2}$$

Rule (3.2) will be referred to *upward consistency rule* or *superset consistency rule.* The proof of this rule follows immediately from the definition of interpretation and the assumption that if $d \in s$, then also $d \in t$. The meaning of the upward consistency rule is obvious — if an attribute takes values from a certain set, then certainly its values stay within any superset of that set.

Third, consider again two subset symbols $s, t \subseteq D$. Obviously, for any object symbol $o \in O$ and any attribute $A_i \in A$, if the attribute takes value in s and in t, then its value must be in $s \cup t$. This observation takes the form of the following rule

$$\frac{A_i(o) = s, A_i(o) = t}{A_i(o) = s \cup t}\,. \tag{3.3}$$

Rule (3.3) will be referred to *union consistency rule.* The proof of this rule follows immediately from the definition of interpretation and the properties of (internal) conjunction. The meaning of this rule is obvious — if an attribute takes values of a certain set and simultaneously of another set, then certainly it takes the values of the sum of them.

Similarly, consider an analogous case of internal disjunction. Consider again two subset symbols $s, t \subseteq D$. Obviously, for any object symbol $o \in O$ and any attribute $A_i \in A$, if the attribute takes value in s and in t, then its value must be in $s \cap t$. This observation takes the form of the following rule

$$\frac{A_i(o) \in s, A_i(o) \in t}{A_i(o) \in s \cap t}\,. \tag{3.4}$$

Rule (3.4) will be referred to *intersection consistency rule.* The proof of this rule follows immediately from the definition of interpretation and the properties of (internal) disjunction. The meaning of this rule is obvious — if an attribute takes values from a certain set and simultaneously belongs to another set, then certainly it takes the values within the intersection of them.

Finally, consider some attribute A_i taking a set t value for certain object o. Then, if t is decomposed into any sets $t_1, t_2, \ldots, t_k$ summing up to t, the value of A_i may be split over these subsets. Hence we have the following rule

$$\frac{A_i(o) = t,\ t = t_1 \cup t_2 \cup \ldots \cup t_k}{A_i(o) = t_1 \wedge A_i(o) = t_2 \wedge \ldots \wedge A_i(o) = t_k}\,. \tag{3.5}$$

The above rule will be referred to as *conjunctive decomposition rule.* The proof of it follows directly from the definition of interpretation. The rule can be applied to decompose a single fact of SAL into a number of facts (in fact — conjunction of them) incorporating smaller subsets of the attribute domain. In case of finite domains, the decomposition may lead to atomic values, i.e. to a formula of AAL.

To end up with this line, consider a certain attribute A_i taking its value for certain object o within set t. Then, if t is decomposed into any sets $t_1, t_2, \ldots, t_k$ summing up to t, the value of A_i must be in one of these subsets.

Hence we have the following rule

$$\frac{A_i(o) \in t,\ t = t_1 \cup t_2 \cup \ldots \cup t_k}{A_i(o) \in t_1 \vee A_i(o) \in t_2 \vee \ldots \vee A_i(o) \in t_k}\,. \tag{3.6}$$

The above rule will be referred to as *disjunctive decomposition rule.* The proof of it follows directly from the definition of interpretation. The rule can be applied to decompose a single fact of SAL into a number of facts (in fact — disjunction of them) incorporating smaller subsets of the attribute domain. In case of finite domains, the decomposition may lead to atomic values, i.e. to a formula of AAL.

4

Resolution

The basic ideas concerning the resolution inference rule and resolution theorem proving were presented in one of the previous chapters. Here we shall refer to some details of the resolution method as applied to first-order logic.

First, let us recapitulate the basic ideas concerning *substitutions* and *unification* of terms and atomic formulae.

4.1 Substitution and Unification

4.1.1 Substitutions

Substitution is an operation allowing to replace some variables occurring in a formula with terms. The goal of applying a substitution is to make a certain formula more specific so that it matches another formula. Typically, substitutions are applied in resolution theorem proving for unification of formulae. A substitution is defined as follows:

Definition 63. *A substitution σ is any finite mapping of variables into terms of the form*

$$\sigma\colon V \to TER.$$

Since substitutions are applied to more complex expressions, it is necessary to extend the definition of substitutions on terms and formulae. This is done in a straightforward way as follows:

Definition 64. *Any substitution σ ($\sigma\colon V \to TER$) is extended to operate on terms and formulae so that a finite mapping of the form*

$$\sigma\colon TER \cup FOR \to TER \cup FOR$$

satisfying the following conditions is induced:

- $\sigma(c) = c$ *for any* $c \in C$;
- $\sigma(X) \in TER$, *and* $\sigma(X) \neq X$ *for a certain finite number of variables only;*

A. Ligêza: *Logical Foundations for Rule-Based Systems*, Studies in Computational Intelligence (SCI) **11**, 65–72 (2006)
www.springerlink.com

- *if* $f(t_1, t_2, \ldots, t_n) \in TER$, *then*

$$\sigma(f(t_1, t_2, \ldots, t_n)) = f(\sigma(t_1), \sigma(t_2), \ldots, \sigma(t_n));$$

- *if* $p(t_1, t_2, \ldots, t_n) \in ATOM$, *then*

$$\sigma(p(t_1, t_2, \ldots, t_n)) = p(\sigma(t_1), \sigma(t_2), \ldots, \sigma(t_n));$$

- $\sigma(\neg\Phi) = \neg(\sigma(\Phi))$, *for any formula* $\Phi \in FOR$;
- $\sigma(\Phi \diamond \Psi) = \sigma(\Phi) \diamond \sigma(\Psi)$ *for any two formulae* $\Phi, \Psi \in FOR$ *and for* $\diamond \in \{\wedge, \vee, \Rightarrow, \Leftrightarrow\}$;
- $\sigma(\nabla X(\Phi)) = \nabla X(\sigma'(\Phi))$ *for any formula* $\Phi \in FOR$ *and* $\nabla \in \{\forall, \exists\}$; *here* $\sigma'(Y) = \sigma(Y)$ *for any* $Y \neq X$ *and* $\sigma'(X) = X$.

Thus, a substitution σ is any finite mapping of variables into terms extended over terms and formulae in the above way. Any formula $\sigma(\Phi)$ resulting from application of substitution σ to the variables of Φ will be denoted as $\Phi\sigma$ and it will be called a *substitution instance* or simply an *instance* of Φ. If no variables occur in $\Phi\sigma$ (or any other formula or term), it will be called a *ground instance* (a *ground formula* or a *ground term*, respectively).

Note, that according to the above definition, substitutions in fact operate only on free variables (they change only free variables, i.e. the ones that are not quantified). For example, in resolution theorem proving all the quantifiers (formally) are removed, and the resulting formulae are quantifier-free; thus all the variables can be regarded as free variables, at least with regard to substitutions application.

Since substitutions operate in fact on a finite number of variables only, they can be conveniently denoted as sets of ordered pairs of variables and the terms to be substituted for them. Hence, any substitution σ can be presented as

$$\sigma = \{X_1/t_1, X_2/t_2, \ldots, X_n/t_n\} ,$$

where t_i is a term to be substituted for variable X_i , $i = 1, 2, \ldots, n$. If Φ is a formula (or term) and σ is a substitution, then $\Phi\sigma$ is the formula (or term) resulting from simultaneous replacing the variables of Φ with the appropriate terms of σ.

Since substitutions are mappings, a *composition* of substitutions is well defined. Note that, having two substitutions, say σ and θ, the composed substitution $\sigma\theta$ can be obtained from σ by *simultaneous* application of θ to all the terms of σ, deletion of any pairs of the form X/t where $t = X$ (identity substitutions), and enclosing all the pairs X/t of θ, such that σ does not substitute for (operate on) X [16, 39].

Let $\sigma = \{X_1/t_1, X_2/t_2, \ldots, X_n/t_n\}$ and let $\theta = \{Y_1/s_1, Y_2/s_2, \ldots, Y_m/s_m\}$. The composition of the above substitutions is obtained from the set

$$\{X_1/t_1\theta, X_2/t_2\theta, \ldots, X_n/t_n\theta, Y_1/s_1, Y_2/s_2, \ldots, Y_m/s_m\}$$

by:

- removing all the pairs $X_i/t_i\theta$ where $X_i = t_i\theta$, and
- removing all the pairs Y_j/s_j where $Y_j \in \{X_1, X_2, \ldots, X_n\}$.

Example. Consider the following substitutions $\sigma = \{X/g(U), Y/f(Z), V/W, Z/c\}$ and $\theta = \{Z/f(U), W/V, U/b\}$. The composition of them is defined as

$$\sigma\theta = \{X/g(b), Y/f(f(U)), Z/c, W/V, U/b\}.$$

Substitutions are in general mappings, but not one-to-one mappings; hence, in general an inverse substitution for a given one may not exist. However, there exists a class of substitutions, the so-called *renaming substitutions*, such that an inverse substitution always exists provided that they are one-to-one mappings.

Definition 65. *Substitution λ is a renaming substitution iff it is off the form*

$$\theta = \{X_1/Y_1, X_2/Y_2, \ldots, X_n/Y_n\} \tag{4.1}$$

Moreover, it is a one-to-one mapping if $Y_i \neq Y_j$ for $i \neq j$, $i, j \in \{1, 2, \ldots, n\}$.

Assume λ is a renaming, one-to-one substitution given by (4.1). The inverse substitution for it is given by $\lambda^{-1} = \{Y_1/X_1, Y_2/X_2, \ldots, Y_n/X_n,\}$. The composition of a renaming substitution and the inverse one leads to an *empty substitution*, traditionally denoted with ϵ; we have $\lambda\lambda^{-1} = \epsilon$.

Let E denote an expression (formula or term), ϵ denote an empty substitution, and let λ be a one-to-one renaming substitution; σ and θ denote any substitution. The following properties are satisfied for any substitutions:

- $E(\sigma\theta) = (E\sigma)\theta$,
- $\sigma(\theta\gamma) = (\sigma\theta)\gamma$ (associativity),
- $E\epsilon = E$,
- $\epsilon\sigma = \sigma\epsilon = \sigma$.

Note that, in general, the composition of substitutions is not commutative.

4.1.2 Unification

Finally, let us move onto unification.

Substitutions are applied to *unify* terms and formulae. Unification is a process of determining and applying a certain substitution to a set of expressions (terms or formulae) in order to make them identical. We have the following definition of unification.

Definition 66. *Let $E_1, E_2, \ldots, E_n \in TER \cup FOR$ are certain expressions. We shall say that expressions $E_1, E_2, \ldots, E_n$ are* unifiable *if and only if there exists a substitution σ, such that $\{E_1, E_2, \ldots, E_n\}\sigma = \{E_1\sigma, E_2\sigma, \ldots, E_n\sigma\}$ is a single-element set.*

Substitution σ satisfying the above condition is called a unifier *(or a* unifying substitution*) for expressions $E_1, E_2, \ldots, E_n$.*

Note that if there exists a unifying substitution for some two or more expressions (terms or formulae), then there usually exists more than one such substitution. It is useful to define the so-called *most general unifier* (*mgu*, for short), which, roughly speaking, substitutes terms for variables only if it is necessary, leaving as much place for possible further substitutions, as possible. The most general unifier is defined as follows.

Definition 67. *A substitution σ is a* most general unifier *for a certain set of expressions if and only if, for any other unifier θ of this set of expressions, there exists a substitution λ, such that $\theta = \sigma\lambda$.*

The meaning of the above definition is obvious. Substitution θ is not a most general unifier, since it is a composition of some simpler substitution σ with an auxiliary substitution λ.

In general, for arbitrary expressions there may exist an infinite number of unifying substitutions. However, it can be proved that any two most general unifiers can differ only with respect to variable names. This is stated with the following theorem.

Theorem 6. *Let θ_1 and θ_2 be two most general unifiers for a certain set of expressions. Then, there exists a one-to-one renaming substitution λ such that $\theta_1 = \theta_2\lambda$ and $\theta_2 = \theta_1\lambda^{-1}$.*

The proof can be found in [16].

As an example consider atomic formulae $p(X, f(Y))$ and $p(Z, f(Z))$. The following substitutions are all most general unifiers:

- $\theta = \{X/U, Y/U, Z/U\}$,
- $\theta_1 = \{Z/X, Y/X\}$,
- $\theta_2 = \{X/Y, Z/Y\}$,
- $\theta_3 = \{X/Z, Y/Z\}$.

All of the above unifiers are equivalent — each of them can be obtained from another one by applying a renaming substitution. For example, $\theta = \theta_1\lambda$ for $\lambda = \{X/U\}$; on the other hand obviously $\theta_1 = \theta\lambda^{-1}$.

4.1.3 Algorithm for Unification

It can be proved that if the analyzed expressions are terms or formulae, then there exists an algorithm for efficient generating the most general unifier, provided that there exists one; in the other case the algorithm terminates after finite number of steps [16]. Thus, the unification problem is decidable.

The basic idea of the unification algorithm can be explained as a subsequent search through the structure of the expressions to be unified for inconsistent relative components and replacing one of them, hopefully being a variable, with the other.

In order to find inconsistent components it is useful to define the so-called disagreement set. Let $W \subseteq TER \cup FOR$ be a set of expressions to be unified.

A *disagreement set* $D(W)$ for a nonempty set W is the set of terms obtained through parallel search of all the expressions of W (from left to right), which are different with respect to the first symbol. Hence, the set $D(W)$ specifies all the inconsistent relative elements met first during the search. The unification algorithm [16] is as follows.

Unification Algorithm

1. Set $i = 0$, $W_i = W$, $\theta_i = \epsilon$.
2. If W_i is a singleton, then stop; θ_i is the most general unifier for W.
3. Find $D(W_i)$.
4. If there are a variable $X \in D(W_i)$ and a term $t \in D(W_i)$, such that X does not occur in t, then proceed; otherwise stop — W is not unifiable.
5. Set $\theta_{i+1} = \theta\{X/t\}$, $W_{i+1} = W_i\{X/t\}$.
6. Set $i = i + 1$ and go to 2.

Example. Consider two atomic formulae $p(X, f(X,Y), g(f(Y,X)))$ and $p(c, Z, g(Z))$. The following steps illustrate the application of the unification algorithm to these atomic formulae.

1. $i = 0$, $W_0 = \{p(X, f(X,Y), g(f(Y,X))), p(c, Z, g(Z))\}$, $\theta_0 = \{\}$.
2. $D(W_0) = \{X, c\}$.
3. $\theta_1 = \{X/c\}$, $W_1 = \{p(c, f(c,Y), g(f(Y,c))), p(c, Z, g(Z))\}$.
4. $D(W_1) = \{f(c,Y), Z\}$.
5. $\theta_2 = \{X/c\}\{Z/f(c,Y)\} = \{X/c, Z/f(c,Y)\}$,
   $W_2 = \{p(c, f(c,Y), g(f(Y,c))), p(c, f(c,Y), g(f(c,Y)))\}$.
6. $D(W_2) = \{Y, c\}$.
7. $\theta_3 = \{X/c, Z/f(c,Y)\}\{Y/c\} = \{X/c, Z/f(c,c), Y/c\}$,
   $W_3 = \{p(c, f(c,c), g(f(c,c))), p(c, f(c,c), g(f(c,c)))\}$.
8. Stop; the most general unifier is $\theta_3 = \{X/c, Z/f(c,c), Y/c\}$.

The unification algorithm has some important properties given by Theorem 7.

Theorem 7. *If W is a finite set of unifiable expressions, then the algorithm always terminates at step 2 and it produces the most general unifier for W. Moreover, if the expressions of W are not unifiable, then the algorithm terminates at step 4.*

The proof can be found in [16].

4.2 Clausal Form

For resolution theorem proving a key issue is to transform the original set of formulae into the so-called *clausal form* i.e. a set of first-order clauses. The procedure is similar to the one applied in case of propositional calculus; in fact one should follow the sequence necessary to transform a formula to CNF. The main difference with respect to propositional logic is that all the quantifiers should also be eliminated.

Apart from the rules coming directly from propositional calculus, there are some new ones applicable to the formulae with quantifiers. Let $\Phi[X]$ mean that variable X occurs in Φ, while Ψ is free from X. These new rules are as follows:

- $\forall X \Phi[X] \wedge \Psi \equiv \forall X(\Phi[X] \wedge \Psi)$,
- $\forall X \Phi[X] \vee \Psi \equiv \forall X(\Phi[X] \vee \Psi)$,
- $\exists X \Phi[X] \wedge \Psi \equiv \exists X(\Phi[X] \wedge \Psi)$,
- $\exists X \Phi[X] \vee \Psi \equiv \exists X(\Phi[X] \vee \Psi)$.

There are also the equivalents to extended De Morgan's laws:

- $\neg(\forall X \Phi[X]) \equiv \exists X(\neg \Phi[X])$,
- $\neg(\exists X \Phi[X]) \equiv \forall X(\neg \Phi[X])$.

Moreover, universal quantifier distributes over conjunction while the existential one — over disjunction:

- $\forall X \Phi[X] \wedge \forall X \Psi[X] \equiv \forall X(\Phi[X] \wedge \Psi[X])$,
- $\exists X \Phi[X] \vee \exists X \Psi[X] \equiv \exists X(\Phi[X] \vee \Psi[X])$.

In the case of universal quantifier and disjunction, as well as in the case of existential quantifier and conjunction, one has to rename variables and proceed according to the following scheme:

- $\forall X \Phi[X] \vee \forall X \Psi[X] \equiv \forall X \Phi[X] \vee \forall Y \Psi[Y] = \forall X \forall Y(\Phi[X] \vee \Psi[Y])$,
- $\exists X \Phi[X] \wedge \exists X \Psi[X] \equiv \exists X \Phi[X] \wedge \exists Y \Psi[Y] = \exists X \exists Y(\Phi[X] \wedge \Psi[Y])$.

The procedure allowing to transform any first-order logic formula to clausal form, necessary for resolution theorem proving consists of subsequent elimination of equivalence and implication connectives, moving negation sign directly before predicate symbols, moving all quantifiers into the front of a formula (forming the so-called *prenex*), and finally, transforming the rest into CNF (the *matrix*). The details are in any textbook on resolution theorem proving, e.g. [16, 37, 39].

4.3 Resolution Rule

Resolution rule in first-order logic constitutes a single and powerful inference rule of conceptually simple scheme, being an extension of the resolution rule in propositional calculus. As in propositional logic it is very attractive, especially for automated theorem proving. Below we explain a single step in resolution theorem proving, i.e. the *resolution rule* and its application.

Let there be given two clauses, $C_1 = \phi \vee q_1$ and $C_2 = \varphi \vee \neg q_2$. It is important that q_1 and $\neg q_2$ are either complementary literals or there exists a most general unifier of the form (mgu) σ, such that $q_1\sigma$ and $q_2\sigma$ are identical, i.e. $q_1\sigma$ and $\neg q_2\sigma$ are complementary. The resolution rule (or resolution principle) allows to generate a new clause $C = \phi\sigma \vee \varphi\sigma$ being a logical consequence of the parent clauses; the complementary literals are removed.

Definition 68. *Let* $C_1 = \phi \vee q_1$ *and* $C_2 = \varphi \vee \neg q_2$ *be two arbitrary clauses. Let* σ *be a mgu for* q_1 *and* q_2. *The Resolution Rule is an inference rule of the form*

$$\frac{\phi \vee q_1, \varphi \vee \neg q_2}{\phi\sigma \vee \varphi\sigma}. \tag{4.2}$$

Obviously, the produced formula is a logical consequence of the parent formulae; the proof is given in almost any handbook on resolution theorem proving, e.g. [16]. Further, note that at the level of propositional language the resolution rule is equivalent to the rule expressing transitivity. In order to see that let us transform the clause to the following equivalent form, i.e. $C_1 = \neg\phi \Rightarrow q_1$ and $C_2 = q_2 \Rightarrow \varphi$. Next, let us apply the mgu σ to both of the formulae; we obtain $C_1\sigma = \neg\phi\sigma \Rightarrow q_1\sigma$ and $C_2\sigma = q_2\sigma \Rightarrow \varphi\sigma$. Now, the resolution rule takes the form

$$\frac{\neg\phi\sigma \Rightarrow q_1\sigma, q_2\sigma \Rightarrow \varphi\sigma}{\neg\phi\sigma \Rightarrow \varphi\sigma}.$$

The resulting formula can be further transformed so we have $\neg\phi\sigma \Rightarrow \varphi\sigma = \phi\sigma \vee \varphi\sigma$.

As in the case of propositional calculus, resolution theorem proving is carried out by appropriate application of the resolution rule. As it was mentioned, the method is especially convenient for automated theorem proving. It has gained a great popularity during the last thirty five years. The resolution method ([16, 37, 39], for precise, logical treatment see as well [144]) combines in a single rule the power of other rules, and due to its uniformity, can be easily implemented for automated theorem proving with computers.

Unfortunately, the resolution rule stated as above does not provide a *complete* tool for refutation. In certain cases it may happen that using the most general unifiers does not lead to proving inconsistency; in such a case one needs the factoring rule.

Let C be any clause such that two or more literals of C can be unified with the most general unifier θ; in this case $C\theta$ is a logical consequence of C ($C \models C\theta$) and $C\theta$ is called a *factor* of C. The rule

$$\frac{C}{C\theta}$$

is called *factorization*. Factorization is a complementary, but necessary rule to assure completeness of resolution theorem proving.

Example. In order to illustrate application of resolution theorem proving let us consider the following example:

> *There is a barber who was ordered to shave anyone who does not shave himself. Should he shave himself or not?*

This is the famous Russell antinomy — in fact the order is inconsistent. Consider the following set of clauses being an equivalent of the problem statement.

A. $\forall X \neg shaves(X, X) \Rightarrow shaves(barber, X)$ — anyone who does not shave himself is shaved by the barber.
B. $\forall Y shaves(barber, Y) \Rightarrow \neg shaves(Y, Y)$ — anyone who is not shaved by the barber shaves himself.

After transmitting to clausal form we have two clauses:

- $C_1 = shaves(X, X) \vee shaves(barber, X)$,
- $C_2 = \neg shaves(barber, Y) \vee \neg shaves(Y, Y)$.

Let us instantiate the variables with substitution $\theta = \{X/barber, Y/barber\}$; we have $C_1\theta = shaves(barber, barber)$ and $C_2\theta = \neg shaves(barber, barber)$. Obviously, there is $C_1 \models C_1\theta$ and $C_2 \models C_2\theta$. Finally, resolution rule can be applied to produce the empty clause as follows

$$\frac{shaves(barber, barber), \neg shaves(barber, barber)}{\bot}.$$

Since the resolvent is a logical consequence of parent clauses, we have $C_1, C_2 \models \bot$, i.e. the initial statement is in fact inconsistent.

To conclude, resolution rule, augmented with factorization rule, constitute a tool for theorem proving which is:

- based on refutation — an empty clause (always false) is to be derived from assumptions completed with negated conclusion,
- sound — any conclusion derived with resolution (and factorization) is sound,
- complete — in the sense that an empty clause can always be deduced from an unsatisfiable set of clauses.

Resolution theorem proving is based on using the clausal form, i.e. quantifier-free first-order CNF formula. Hence it is especially convenient for systems which are or can be easily transformed into CNF. In rule-based systems resolution theorem proving finds the following applications:

- proving satisfaction of preconditions of rules in order to check if a selected rule can be fired (see [142]),
- proving attainability of goals,
- checking for inconsistent rules.

Resolution is also the basic rule implemented in all PROLOG systems.

5

Dual Resolution

Dual resolution, or more precisely, *backward dual resolution* (*bd-resolution*, for short) is an inference method dual to classical resolution. An interesting fact is that this rule works backwards in the sense that the disjunction of the parent formulae is a logical consequence of the inferred formula. The logical foundations for dual resolution at the level of propositional logic were presented in Chap. 1 in Subsect. 1.8.3. Here dual resolution in first-order logic is presented in detail.

5.1 Minterm Form

For theorem proving with bd-resolution a key issue is to transform the original set of formulae into the so-called *minterm form* i.e. a set of first-order, quantifier-free minterms (simple conjunctive formulae). The procedure is analogous to the one applied in case of transforming a formula to DNF in propositional calculus; in fact one should follow the sequence necessary to transform a formula to DNF. The main difference with respect to propositional logic is that all the quantifiers should also be eliminated.

Apart from the rules coming directly from propositional calculus, there are some new ones applicable to the formulae with quantifiers. These rules are the same as in case of resolution, see Sect. 4.2.

The procedure allowing to transform any first-order logic formula to minterm form, necessary for bd-resolution theorem proving is composed of subsequent steps aimed at elimination of equivalence and implication connectives, moving negation sign directly before predicate symbols, moving all quantifiers into the front of a formula (forming the so-called *prenex*), and finally, transforming the rest into DNF (the *matrix*). The details are the same as in the case of resolution and one can find them in any textbook on resolution theorem proving, e.g. [16]. There are, however, two differences.

The first difference is that the matrix of the formula is to be transformed into the DNF form rather than CNF. This may be inconvenient in case of large

A. Ligêza: *Logical Foundations for Rule-Based Systems*, Studies in Computational Intelligence (SCI) **11**, 73–88 (2006)
www.springerlink.com

sets of axioms which form a conjunction of statements from logical point of view. Thus, bd-resolution theorem proving may be recommended either if the initial form is 'close to' DNF, or for proving completeness (tautology) of a set of formulae. The latter case is for example the case of proving completeness of rule-based systems.

The second difference concerns elimination of universal quantifiers rather than the existential ones. This can be done in a way analogous to skolemization applied in classical resolution.

Let $Q_1X_1\ Q_2X_2\ \ldots\ Q_nX_n\Psi$ be the prenex normal form obtained from the initial formula, where Q_i, $i = 1, 2, \ldots, n$ are all the quantifiers and Ψ is the quantifier-free matrix of the formula in DNF. Assume that Q_i is the first universal quantifier encountered when scanning the prefix of the formula from left to the right. Now there are two possibilities:

1) If no existential quantifier occurs before Q_i then all the occurrences of variable X_i in Ψ are replaced with a new constant c (c cannot occur in Ψ) and Q_iX_i is removed from the prefix.
2) If $Q_{k_1}, Q_{k_2}, \ldots, Q_{k_j}$ are all the existential quantifiers occurring before Q_i, then all the occurrences of X_i in Ψ are replaced with a term of the form $f(X_{k_1}, X_{k_2}, \ldots, X_{k_j})$, where f is a new function symbol, and Q_iX_i is removed from the prefix.

In this way all the universal quantifiers are eliminated from the prefix, and, since all the variables are existentially quantified, the prefix can be omitted. From now on it is assumed that all the variables in the resulting minterm form (DNF) are implicitly existentially quantified. It is important to note, that although the quantifier-free formula obtained in this way is not necessarily logically equivalent to the initial formula, they are simultaneously tautological formulae.

Theorem 8. *Let Ω be a logical formula and let Φ be its quantifier-free minterm form. Ω is a tautology if and only if Φ is a tautology.*

Proof. Let

$$\Omega = Q_1X_1\ Q_2X_2\ \ldots\ Q_nX_n\Psi\ .$$

Let us scan the prefix of Ω from left to right until the first universal quantifier is encountered; let it be at position i and the encountered symbol is $Q_iX_i = \forall X_i$. In order to eliminate this quantifier and variable X_i we remove the symbols Q_iX_i from the prefix and substitute $f(X_1, \ldots, X_{i-1})$ for any occurrence of X_i in Ψ — in this way we obtain a new formula Ω_i, where

$$\Omega_i = \exists X_1\ \ldots\ \exists X_{i-1}, Q_{i+1}X_{i+1}, \ldots, Q_nX_n\Psi\{X_i/f(X_1, \ldots, X_{i-1})\}\ .$$

The main point is to prove that Ω is tautology iff Ω_i is tautology.

Assume first that Ω is tautology. Thus, for any interpretation I, there exist values of $X_1, \ldots, X_{i-1}$ such that *for any* value of X_i formula $Q_{i+1}X_{i+1}$, $\ldots, Q_nX_n\Psi$ is tautology. By hypothesis, assume that Ω_i is not tautology.

Hence, there exists an interpretation I' falsifying it. In other words, there exist values of $X_1, \ldots, X_{i-1}$ and a value of $f(X_1, \ldots, X_{i-1})$ such that formula $Q_{i+1}X_{i+1}, \ldots, Q_nX_n\Psi$ is false. Obviously, this hypothesis is inconsistent with the assumption that Ω is tautology, since for $X_i = f(X_1, \ldots, X_{i-1})$ it would be false.

Now, assume that Ω_i is tautology. Thus for any interpretation I, there exist values of $X_1, \ldots, X_{i-1}$ such that $Q_{i+1}X_{i+1}, \ldots, Q_nX_n\Psi\{X_i/f(X_1, \ldots, X_{i-1})\}$ is true. By hypothesis, assume that Ω is not a tautology. Hence, there exists an interpretation I' and some values of $X_1, \ldots, X_{i-1}$ as well as a value of X_i such that $Q_{i+1}X_{i+1}, \ldots, Q_nX_n\Psi$ is false. Obviously, this leads to inconsistency, since Ω_i is true under *any* interpretation of $f(X_1, \ldots, X_{i-1})$.

The above line of reasoning should be applied inductively to eliminate all universal quantifiers, when scanning the prefix of Ω for increasing values of i. □

The idea of the above proof is based on the proof of Theorem 4.1 concerning properties of classical Skolemization for resolution theorem proving, as presented in [16]. A proof of this theorem can also be found in [7].

5.2 Introduction to Dual Resolution

Consider for intuition the following simple example from [55]. Let there be given a formula ψ, $\psi = p(a) \wedge q(c) \wedge r(b)$ and another three formulae $\phi_1 = p(a) \wedge q(b)$, $\phi_2 = p(X) \wedge \neg q(X) \wedge r(X)$ and $\phi_3 = \neg p(Y) \wedge r(Y)$ (recall that all the variables are denoted with capitals and are assumed to be implicitly existentially quantified). The problem is to check if the disjunction of ϕ_1, ϕ_2 and ϕ_3 'covers' ψ, i.e. if any world satisfying ψ satisfies the disjunction of ϕ_1, ϕ_2 and ϕ_3. In other words, we check if the disjunction constitutes a formula which is logical consequence of ψ. With use of the proposed approach we may attempt to perform this check as follows.

Let us substitute b for X in ϕ_2; ϕ_2 is logical consequence of the resulting formula; applying such a substitution will be called *factorization*. Further, we can combine (analogically to resolution) the resulting formula with ϕ_1 — by 'resolving' upon the pair $q(b)$, $\neg q(b)$ we obtain formula $\phi' = p(a) \wedge p(b) \wedge r(b)$ and the initial disjunction of ϕ_1 and ϕ_2 is logical consequence of ϕ'; note that under assumption that ϕ' is satisfied (all the components $p(a)$, $p(b)$ and $r(b)$ must be satisfied), the disjunction of ϕ_1 and ϕ_2 with b substituted for X must be satisfied as well — this follows from the 'key' observation that from the two literals $q(b)$ and $\neg q(b)$ at least one must be satisfied[1]. Similarly, we can combine ϕ' with ϕ_3 after substituting b for Y; we obtain $\phi'' = p(a) \wedge r(b)$ and

[1] In fact, in classical logic — and so is our case — exactly one of them is satisfied; however, one can develop further extensions of the discussed reasoning principle for which it would be enough to have *at least one* of the literals (formulae) satisfied. This is so in case of the generalized bd-resolution presented in [53, 55, 56].

the disjunction of ϕ_3 and ϕ' is logical consequence of ϕ''. On the other hand ϕ'' is more general than ψ, thus the disjunction of initial formulae ϕ_1, ϕ_2 and ϕ_3 really 'covers' ψ.

Now let us consider the general case of bd-resolution. Let Ψ be a normal formula, $\Psi = \psi_1 \vee \psi_2 \vee \ldots \vee \psi_n$, where ψ_k is a simple formula for $k = 1, 2, \ldots, n$. Using the set notation we can also write $[\Psi] = \{\psi_1, \psi_2, \ldots, \psi_n\}$. Further, let ψ_i and ψ_j be some two simple formulae of Ψ, such that $\psi_i = \psi'_i \wedge p_i$, and $\psi_j = \psi'_j \wedge \neg p_j$, where p_i and p_j are atomic formulae. If p_i and p_j are unifiable, i.e. if there exists a substitution σ, such that $p_i\sigma = p_j\sigma$, then it is possible to build a new simple formula by combining together ψ_i and ψ_j.

In order to do that let us first apply σ to both ψ_i and ψ_j; we obtain $\psi_i\sigma = \psi'_i\sigma \wedge p_i\sigma$ and $\psi_j\sigma = \psi'_j\sigma \wedge \neg p_j\sigma$. Note that applying the substitution to ψ_i and ψ_j cannot increase their generality. Note also, that after applying the unifying substitution, there is $p_i\sigma = p_j\sigma$, i.e. we have arrived at a complementary pair of literals within the above simple formulae. Now, since either $p_i\sigma$ or $\neg p_j\sigma$ must be false (under any interpretation), one can replace the disjunction of $(\psi'_i\sigma \wedge p_i\sigma) \vee (\psi'_j\sigma \wedge \neg p_j\sigma)$ with a conjunction of the form $\psi'_i\sigma \wedge \psi'_j\sigma$.

It can be seen that such a replacement cannot increase generality of the initial formula, so the initial formula is logical consequence of the derived one. This is so, since if the resulting formula is satisfied under a certain interpretation, one of the simple formulae constituting the initial disjunction must be satisfied as well (which one depends on the interpretation and the truth-value of $p_i\sigma$ under it). Thus, one can expect that the initial formula is at least as general as the generated one, to be called a *bd-resolvent*.

In fact, starting from Ψ one can repeat the process in a recursive way, and at any stage the initial formula is at least as general as the last bd-resolvent generated, so the logical consequence relation is kept in the reversed order with respect to the one of generation of new formulae.

If, after a finite number of steps one arrives at an *empty* formula ($\top$, always true, generated from a disjunction of the form $p \vee \neg p$), one can conclude that the initial formula is satisfied under any interpretation, and as such is a valid formula. This observation gives rise to a formal method of theorem proving.

5.3 Dual Resolution Rule

As the classical resolution rule, the dual resolution rule in first-order logic constitutes a single and powerful inference rule of conceptually simple scheme. Simultaneously, it constitutes an extension of the bd-resolution rule in propositional calculus, see Subsect. 1.8.3. As in propositional logic it is very attractive, especially for direct proving that a formula is tautology, in the case of formulae in DNF form.

Below we explain a single step in bd-resolution theorem proving, i.e. the *bd-resolution rule* and its application.

Let there be given two minterms, $M_1 = \phi \wedge q_1$ and $M_2 = \varphi \wedge \neg q_2$. It is important that q_1 and $\neg q_2$ are either complementary literals or there exists a most general unifier σ, such that $q_1\sigma$ and $q_2\sigma$ are identical, an so $q_1\sigma$ and $\neg q_2\sigma$ are complementary. The bd-resolution rule (or bd-resolution principle) allows to generate a new simple formula $M = \phi\sigma \wedge \varphi\sigma$; the complementary literals are removed.

A graphical presentation is given below (Fig. 5.1).

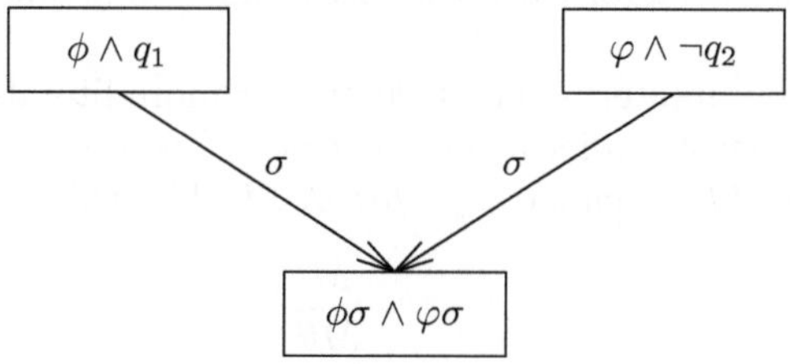

Fig. 5.1. A schematic presentation of dual resolution

What is important and constitutes a principal difference with respect to classical resolution rule is that the disjunction of the parent minterms is a logical consequence of the generated result; this is schematically presented below (Fig. 5.2).

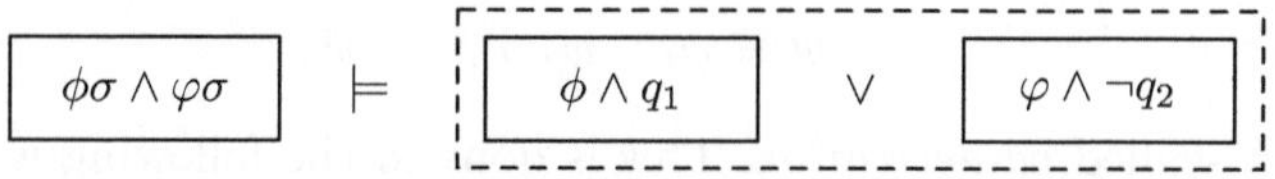

Fig. 5.2. Logical consequence in dual resolution

The basic rule of backward dual resolution is defined as follows.

Definition 69 (Dual Resolution Rule). *Let $M_1 = \phi \wedge q_1$ and $M_2 = \varphi \wedge \neg q_2$ be two arbitrary minterms. Let σ be a mgu for q_1 and q_2. The Backward Dual Resolution Rule is an inference rule of the form*

$$\frac{\phi \wedge q_1, \varphi \wedge \neg q_2}{\phi\sigma \wedge \varphi\sigma}\,. \tag{5.1}$$

Obviously, the produced formula is not a logical consequence of the parent formulae. In a sense the rule works backwards — the disjunction of the parent minterms is a logical consequence of the result, i.e. there is $M \models M_1 \vee M_2$.

Note that at the level of propositional language the bd-resolution rule is 'syntactically similar' to the so-called *Consensus Rule*. The rule is of the form

$$\frac{p \wedge q, \neg q \wedge r}{p \wedge r}\,.$$

However, as mentioned above, the consensus rule stated as above is not a valid inference rule.

As in the case of propositional calculus, bd-resolution theorem proving is carried out by appropriate application of the bd-resolution rule. It seems natural that the method is especially convenient for automated theorem proving, especially in the case the initial formula is in DNF. An example of practical applications includes proving of completeness of rule-based systems.

By analogy to classical resolution theorem proving, in theorem proving with bd-resolution factorization is also a necessary additional rule to assure completeness.

Let M be any minterm such that two or more literals of M can be unified with a most general unifier θ; in this case M is a logical consequence of $M\theta$ ($M\theta \models M$) and $M\theta$ is called a *factor* of M. The rule

$$\frac{M}{M\theta}$$

is called *factorization*. Factorization is a complementary, but necessary rule to assure completeness of bd-resolution theorem proving.

5.4 BD-Derivation

Now, let us define the way in which one can generate a sequence of bd-resolvents starting from an initial normal formula

$$\Psi = \psi_1 \vee \psi_2 \vee \ldots \vee \psi_m \tag{5.2}$$

i.e. the so-called *bd-derivation*. This is done in the following way.

Definition 70. *A* bd-derivation *(or derivation, for short) of a simple formula ψ from a normal formula Ψ given by* (5.2) *is any sequence of simple formulae $\psi^1, \psi^2, \ldots, \psi^k$, such that:*

- *for any $j \in \{1, 2, \ldots, k\}$ ψ^j is either a factor of some ψ^i or a bd-resolvent of simple formulae ψ^i, $\psi^{i'}$, where either $i \leq j$ or $\psi^i \in \Psi$ and $i' \leq j$ or $\psi^{i'} \in \Psi$;*
- $\psi = \psi^k$.

Formula ψ is said to be bd-derived *from Ψ.*

A formula ψ can be derived from some normal formula Ψ by generating a sequence of simple formulae, such that any formula in the sequence is either a factor of, or a bd-resolvent of some earlier generated formulae (or the ones in Ψ); any formula in the sequence is said to be bd-derived from Ψ, and if ψ appears as the last formula in the above sequence, then it is bd-derived from Ψ as well. This will be denoted shortly as $\Psi \vdash_{DR} \psi$. For simplicity, in case of no ambiguity, we shall also say that ψ is derived from Ψ and we shall write $\Psi \vdash \psi$.

Example. In order to illustrate application of bd-resolution theorem proving let us consider the following example extracted from [111].

As the original example is not a single-layered system, we present below a 'flattened' version, equivalent with respect to completeness checking. There are the following three rules for deciding about the STATUS of some person:

$$\begin{aligned}
\mathtt{UNIV_MEMBER}(X) \wedge \mathtt{ENROLLED}(X) \wedge \mathtt{HAS_BS_DEGREE}(X) &\rightarrow \mathtt{STATUS}(X, \mathit{graduateStudent});\\
\mathtt{UNIV_MEMBER}(X) \wedge \mathtt{ENROLLED}(X) \wedge \neg\mathtt{HAS_BS_DEGREE}(X) &\rightarrow \mathtt{STATUS}(X, \mathit{undergraduate});\\
\mathtt{UNIV_MEMBER}(X) \wedge \neg\mathtt{ENROLLED}(X) \wedge \mathtt{HAS_BS_DEGREE}(X) &\rightarrow \mathtt{STATUS}(X, \mathit{staff});\\
\neg\mathtt{ENROLLED}(X) \wedge \neg\mathtt{HAS_BS_DEGREE}(X) &\rightarrow \mathtt{STATUS}(X, \mathit{nonAcademic});\\
\neg\mathtt{UNIV_MEMBER}(X) &\rightarrow \mathtt{STATUS}(X, \mathit{nonAcademic}).
\end{aligned}$$

Using the bd-resolution one can produce most general formulae specifying logical completeness of preconditions of the rules, i.e. showing that in *any* case of input data at least one rule can be fired (as it covers the case). The proof goes as follows.

By taking the preconditions of the first and second rules, one can generate their bd-resolvent of the form

$$\psi_1 = \mathtt{UNIV\_MEMBER}(X) \wedge \mathtt{ENROLLED}(X).$$

Similarly, taking the preconditions of first and third rules one can generate bd-resolvent of the form

$$\psi_2 = \mathtt{UNIV\_MEMBER}(X) \wedge \mathtt{HAS\_BS\_DEGREE}(X).$$

Note that $\psi_1 \vee \psi_2$ specifies the *positive* cases covered by the system (academic persons). The system is *specifically logically complete* with respect to $\psi = \psi_1 \vee \psi_2$.

Now, apply the non-academic cases specification given by rules four and five. For intuition, the above rules cover any non UNIV_MEMBER nor anyone not ENROLLED and such that HAS_BS_DEGREE is not satisfied.

By applying bd-resolution to ψ_1 and precondition of the fourth rule one obtains $\mathtt{UNIV\_MEMBER}(X) \wedge \neg\mathtt{HAS\_BS\_DEGREE}(X)$, and by further bd-resolving with ψ_2 one obtains $\mathtt{UNIV\_MEMBER}(X)$. Finally, after resolving the result with the preconditions of the fifth rule one obtains the empty formula $\top$ (always true). Hence, the disjunction of the preconditions of the above rules is tautology.

5.5 Properties of BD-Resolution

The presented inference method consists in automated deduction by bd-resolution; it is analogous to the well known *resolution method* (for resolution see e.g. [16, 37, 39]. Although the literature concerning resolution is abundant, very little attention was paid to its dual version. The possibilities of using such a dual method are just mentioned in some books; in [7] (Chapter IV.1) a common definition of resolution and its dual version are presented. Further, in [8] a note on *Consolution* method following from *Connection Graphs* is presented. However, it seems that the method of dual resolution was largely ignored in the literature.

The presented reasoning method is similar to resolution, since a basic step of reasoning is accomplished by 'combining' two selected formulae in order to generate some outcome formula. Further, bd-resolution operates on standardized formulae, namely the normal ones, and when used for theorem proving, the goal is to arrive at an 'empty' formula. BD-resolution constitutes, in fact, a single inference rule which is both *sound* and *complete.*

The discussed inference rule is in fact *dual* to the resolution rule — there is a direct one-to-one mapping between *clauses* in resolution theorem proving and *minterms* or *simple formulae* in bd-resolution, as well as between a *resolvent* obtained from two clauses, and a *bd-resolvent* obtained from the respective normal formulae. The mapping, roughly speaking, is based on replacing disjunction with conjunction and positive literals with respective negative ones, and vice versa. Simultaneously, bd-resolution works, in a sense, *backwards*, since in fact the input formula being a disjunction of simple formulae is the logical consequence of the generated bd-resolvent, i.e. the direction of *logical inference* (*taking logical consequences*) is opposite to the one of generating new formulae. Thus, it seems that the term *backward dual resolution* is very close to the idea of the proposed method.

5.5.1 Soundness of BD-Resolution

Soundness of bd-resolution means, that whenever a formal bd-derivation of a formula is found, then the disjunction of initial minterm formulae is logical consequence of the derived formula.

First, let us present the following lemma.

Lemma 8. *Let Φ be any formula (in minterm form) and let σ denote any substitution. Then, if $\models_I \Phi\sigma$, then, there is also $\models_I \Phi$.*

Proof. Assume I is an interpretation satisfying the instance of Φ, i.e. there is $\models_I \Phi\sigma$. Assume σ is specified as $\sigma = \{X_1/t_1, \ldots, X_n/t_n\}$. Let U denote the universe for I. Assume that I does not satisfy Φ; in such a case there would not exist any mapping of variables of Φ into U such that Φ is satisfied. However, such a mapping exists for $\Phi\sigma$, which is contradictory — it would be enough to assign to $X_1, \ldots, X_n$ the values to which terms $t_1, \ldots, t_n$ are mapped with I. □

The following theorem assures soundness of bd-resolution.

Theorem 9 (Soundness of bd-resolution rule). *Let Ψ be a formula of the form* (5.2) *and let $\psi_i = \phi \wedge q_1$ and $\psi_j = \varphi \wedge \neg q_2$ be any two minterms of Ψ. Moreover, let $\psi = \phi\sigma \wedge \varphi\sigma$ be the bd-resolvent of them. Then*

$$\psi \models \psi_i \vee \psi_j \tag{5.3}$$

and $\psi \models \Psi$.

Proof. First let us notice that for the formula Ψ, defined by (5.2) it is true that $\psi_i \vee \psi_j \models \Psi$. Thus, with regard to transitivity of logical consequence, it is enough to prove that $\psi \models \psi_i \vee \psi_j$.

Now, let us assume that ψ is satisfied under some interpretation I, *i.e.* that $\models_I \psi$. Then, obviously, $\models_I \phi\sigma$ and $\models_I \varphi\sigma$, since ψ is the conjunction of them. Further, for the complementary literals $q_1\sigma$ and $\neg q_2\sigma$ there is either (i) $\models_I q_1\sigma$ and $\not\models_I \neg q_2\sigma$ or (ii) $\not\models_I q_1\sigma$ and $\models_I \neg q_2\sigma$. In case (i) there is $\models_I \phi\sigma \wedge q_1\sigma$; in case (ii) we have $\models_I \varphi\sigma \wedge \neg q_2\sigma$. Hence, no matter which is the case, there is $\models_I (\phi\sigma \wedge q_1\sigma) \vee (\varphi\sigma \wedge \neg q_2\sigma)$, i.e. $\models_I (\psi_i \vee \psi_j)\sigma$. Further, by lemma 8, $\models_I \psi_i \vee \psi_j$. □

Now let us return to bd-derivation. The following theorem assures soundness of bd-derivation.

Theorem 10 (Soundness of bd-derivation). *Let Ψ be a formula of the form* (5.2) *and let ψ be a simple formula obtained by bd-derivation from Ψ. Then $\psi \models \Psi$.*

Proof. Let $\psi^1, \psi^2, \ldots, \psi^k$ be the bd-derivation of ψ from Ψ. We have $\psi = \psi^k$ and any of the formulae ψ^i is a bd-resolvent or a factor of earlier generated formulae (or the ones belonging to Ψ), $i = 1, 2, \ldots, k$.

Let us build a sequence of normal formulae $\Psi^0, \Psi^1, \Psi^2, \ldots, \Psi^k$, such that $\Psi^0 = \Psi$, $\Psi^1 = \Psi^0 \vee \psi^1$, $\Psi^2 = \Psi^1 \vee \psi^2, \ldots, \Psi^i = \Psi^{i-1} \vee \psi^i, \ldots, \Psi^k = \Psi^{k-1} \vee \psi^k$, i.e. at any stage we adjoin the newly generated formula by use of disjunction. By the above theorem on soundness of bd-resolution rule we have $\psi_i \models \Psi_{i-1}$, since ψ_i is a bd-resolvent of minterms from Ψ_{i-1}. Hence, obviously, for any two formulae $\Psi^i = \psi_i \vee \Psi^{i-1}$ and Ψ^{i-1} in the above sequence we have $\Psi^i \models \Psi^{i-1}$ for $i = 1, 2, \ldots, k$; in fact, at any stage we join the current formula with a less or equally general bd-resolvent. Further, $\psi \models \Psi^k$. Thus, with regard to transitivity of logical consequence we conclude that $\psi \models \Psi$. □

5.5.2 Completeness of BD-Resolution

BD-resolution rule for theorem proving, together with the factorization rule, constitute a complete system of rules for theorem proving in first-order logic. More precisely, the rules are complete with respect to proving logical completeness of a (disjunctive) set of minterms, i.e. a formula in DNF. This means that whenever such a formula is tautology, a derivation of an empty formula (one always true) with bd-resolution exists. Moreover, and this is an even stronger result, bd-resolution is complete with respect to proving logical consequence. This means that whenever a simple formula ϕ is a logical consequence of a normal formula Ψ, then it is always possible to prove it using bd-resolution.

In order to prove the general theorem concerning completeness of bd-resolution, let us first prove its initial version concerning completeness of bd-resolution in case of *ground formulae*, i.e. ones with no variables.

Theorem 11 (Ground Completeness Theorem). *Let Ψ be any ground normal formula, and let ϕ be a simple formula. If*

$$\phi \models \Psi \tag{5.4}$$

then there exists a bd-derivation of a ground simple formula ψ from Ψ, such that

$$\phi \models \psi \,. \tag{5.5}$$

Moreover, if ϕ is satisfiable, then also

$$\psi \subseteq \phi \,, \tag{5.6}$$

i.e. the derived formula ψ subsumes (with empty substitution) ϕ.

Proof. If ϕ is unsatisfiable, then any bd-derivation satisfies the theorem; relation (5.4) is trivially satisfied. Assume that ϕ is satisfiable (i.e. ϕ does not contain any pair of complementary literals).

After [39][2], let us define the *number of excess literals* in Ψ as the number of all literal occurrences minus the number of minterms (simple formulae) in Ψ. Obviously, for n being the number of excess literals in any Ψ, there is $n \geq 0$; $n = 0$ means that all the simple formulae are just literals. The proof is done by induction with regard to n.

Consider the case when $n = 0$ (all the simple formulae of Ψ are just ground literals). Since $\phi \models \Psi$, then either Ψ must contain at least one simple formula — being in fact a ground literal q — such that $q \in [\phi]$, or Ψ must contain a pair of ground complementary literals which bd-resolve giving the empty formula $\top$ (always true); in either case the necessary bd-derivation producing ψ exists — in the former one it is just $\psi = q$, while in the latter one the bd-resolution of these two literals leading to the empty formula constitutes the bd-derivation of interests.

Now, assume that the above theorem is true for all ground normal formulae having less than n, $n \geq 1$, excess literals. Let Ψ contain n excess literals. Thus Ψ must contain at least one simple formula, say φ, having more than one literal.

Let us select an arbitrary literal, say q, belonging to φ, and let us define $\varphi_q = \varphi \setminus \{q\}$. Note that, since $\varphi \models \varphi_q$, there is also $\Psi \models (\Psi \setminus \varphi) \cup \varphi_q$. Further, since $(\Psi \setminus \varphi) \cup \varphi_q$ contains one less excess literal (i.e. $n - 1$), it must satisfy the theorem, i.e. if $\phi \models (\Psi \setminus \varphi) \cup \varphi_q$, then there exists a bd-derivation of some ψ_φ satisfying the theorem from it; let us label this bd-derivation $bdd\colon \psi_\varphi$.

[2] This and the following three proofs concerning completeness of bd-resolution are based on the ideas of the appropriate proofs of completeness of resolution presented in [39].

On the other hand, $\Psi \models (\Psi \setminus \varphi) \cup \{q\}$, and if $\phi \models (\Psi \setminus \varphi) \cup \{q\}$, there also must exist a bd-derivation of some ψ_q satisfying the theorem from $(\Psi \setminus \varphi) \cup \{q\}$; let us label this bd-derivation *bdd*: ψ_q.

Now, if φ_q is not used for obtaining the former bd-derivation (*bdd*: ψ_φ), the derivation of ψ from Ψ exists; in fact, $\psi = \psi_\varphi$. If not, one can construct the bd-derivation as follows. Add q back to φ_q and all its descendants in the bd-derivation. If an appropriate formula ψ is generated (either the empty one or not), then the bd-derivation of interest exists. Otherwise, we arrive at a ground simple formula consisting of a conjunction of ψ_φ (which itself satisfies the theorem) and the single literal q, i.e. we stop at $\psi_\varphi \wedge q$. But now, one can append the bd-derivation of the other simple formula (*bdd*: ψ_q), i.e. ψ_q starting from $(\Psi \setminus \varphi) \cup (q \wedge \psi_\varphi)$. Thus, the bd-derivation of a simple formula satisfying the theorem exists, and finally $\psi = \psi_q \wedge \psi_\varphi$. □

From the above theorem one can learn that for the case of Ψ being a ground normal formula the bd-resolution is complete in the wider sense. Note, that we need not assume that ϕ is a ground formula; however, it can be seen, that if ϕ is a satisfiable formula and Ψ is not a tautology, then at least a part of ϕ must be ground (to cover the literals of ψ).

In the following part we shall attempt to generalize the result for Ψ being any normal formula (not necessarily a ground one).

First, let us now restate the so-called *lifting lemma* [39] in a suitable form.

Lemma 9 (Lifting Lemma). *Let ψ and ϕ be two simple formulae with no variables in common (i.e. $FV(\psi) \cap FV(\phi) = \emptyset$), and let ψ', ϕ' be some ground simple formulae, such that $\psi' = \psi\theta$ and $\phi' = \phi\theta$ for some (ground) substitution θ. If φ' is a bd-resolvent of ψ' and ϕ', then there exists a bd-resolvent φ of ψ and ϕ and a substitution λ, such that $\varphi' = \varphi\lambda$, i.e. φ' is a substitution instance of φ.*

Proof. Since φ' is a bd-resolvent of ψ' and ϕ', then there must exist a pair of complementary ground literals, say q' and $\neg q'$, such that $q' \in \psi'$ and $\neg q' \in \phi'$. Further, $\varphi' = (\psi' \setminus \{q'\}) \cup (\phi' \setminus \{\neg q'\})$.

Let $\{p_1, p_2, \ldots, p_m\}$ be all the literals of ψ which are mapped by θ to q'. Similarly, let $\{\neg q_1, \neg q_2, \ldots, \neg q_n\}$ be the set of all the literals of ϕ which are mapped by θ to $\neg q'$. Let σ_p be the most general unifier for the first set of literals, and let σ_q be the most general unifier for the second one. Further, let $\sigma = \sigma_p \cup \sigma_q$ be the composite substitution. We put $p_\sigma = p_i\sigma$, $i = 1, 2, \ldots, m$, and $q_\sigma = q_j\sigma$, $j = 1, 2, \ldots, n$.

From the definition of the most general unifier it follows that q' must be an instance of p_σ and simultaneously q' must be an instance of q_σ; thus a unifier for p_σ and q_σ exists. Let γ be the most general unifier for p_σ and q_σ. Now let us define the bd-resolvent of ψ and ϕ as follows

$$\varphi = (\psi\sigma\gamma \setminus \{p_1, p_2, \ldots, p_m\}\sigma\gamma) \cup (\phi\sigma\gamma \setminus \{\neg q_1, \neg q_2, \ldots, \neg q_n\}\sigma\gamma) .$$

Note, that with regard to the defined substitutions, φ' can be written as follows

$$\varphi' = (\psi\theta \setminus \{p_1, p_2, \ldots, p_m\}\theta) \cup (\phi\theta \setminus \{\neg q_1, \neg q_2, \ldots, \neg q_n\}\theta) .$$

Since q' is an instance of p_σ and q_σ, θ is a substitution composed of $\sigma\gamma$ with some other substitution. Hence, φ' is an instance of φ. □

The above *lifting lemma* can be further generalized from a single bd-resolvent case to bd-derivation, so that we obtain the so-called *lifting theorem.*

Theorem 12 (Lifting Theorem). *Let Ψ be a normal formula and let Ψ' be a ground normal formula, such that $\Psi\sigma = \Psi'$ for some substitution σ. If there exists a bd-derivation of a simple formula ψ' from Ψ', then there exists a bd-derivation of a simple formula ψ from Ψ, such that ψ' is a substitution instance of ψ (i.e. for some substitution σ', there is $\psi\sigma' = \psi'$).*

Proof. Any bd-derivation consists of a finite number of bd-resolution steps, and any step is based on either bd-resolution or factorization (application of a substitution). Since Ψ' is a ground formula, in fact no factorization takes place. The case of bd-resolution is dealt with by use of the lifting lemma (proved above). Thus, by simple induction with regard to the length of the bd-derivation the proof of the theorem is straightforward. □

The final theorem assuring completeness of bd-resolution can be stated as follows.

Theorem 13 (Completeness Theorem). *Let Ψ be any normal formula, and let ϕ be some simple formula. Assume that Ψ and ϕ have no variables in common, i.e. $FV(\Psi) \cap FV(\phi) = \emptyset$[3]. If*

$$\phi \models \Psi \tag{5.7}$$

then there exists a bd-derivation of a simple formula ψ from Ψ, such that

$$\phi \models \psi . \tag{5.8}$$

Moreover, if ϕ is satisfiable, then there exists a substitution θ such that

$$[\psi\theta] \subseteq [\phi] , \tag{5.9}$$

i.e. the derived formula ψ subsumes ϕ.

Proof. In case ϕ is unsatisfiable, the theorem is trivially satisfied. Assume ϕ is a satisfiable simple formula.

Assume that (5.7) holds, i.e. $\phi \models \Psi$. Let σ be an arbitrary ground, one-to-one substitution, such that $FV(\phi\sigma) = \emptyset$ and $C(\sigma) \cap [C(\Psi) \cup C(\phi)] = \emptyset$, i.e.

[3] If not, a simple renaming of variables may be necessary.

σ replaces all the variables of ϕ with new constants, not occurring in Ψ or ϕ. Let $\phi\sigma = \phi_\sigma$. Of course, $\phi_\sigma \models \Psi$.

Now, by the version of Herbrand Theorem (Theorem 5 presented in Sect. 2.5), if $\phi_\sigma \models \Psi$, then there exists a ground substitution instance Ψ' of Ψ, such that $\phi_\sigma \models \Psi'$. Further, by the Ground Completeness Theorem, there exists a bd-derivation of ψ' from Ψ', such that $\phi_\sigma \models \psi'$ and also $\psi' \subseteq \phi_\sigma$. Now, by the lifting theorem, there exists a bd-derivation of some formula ψ_σ from Ψ, such that $\psi' \models \psi_\sigma$ and ψ' is a substitution instance of ψ_σ. Hence $\phi_\sigma \models \psi_\sigma$. What we have proved is that for any σ satisfying the above assumption, there exists a bd-derivation of some simple formula ψ_σ from Ψ, such that $\phi_\sigma \models \psi_\sigma$ and ϕ_σ is a substitution instance of ψ_σ, i.e. for some substitution θ' there is $\psi_\sigma\theta' \subseteq \phi_\sigma$.

Let us notice that since Ψ and ϕ do not have variables in common, it is possible to 'mechanically' replace the constants introduced to ϕ by σ with the original variables, and repeat the bd-derivation keeping the variables unchanged (i.e. if any of them occur in the derivation, no substitution can be applied to alter these variables; this is possible, since the initial derivation was performed with constants). Note that the required this time ψ (see (5.8)) can be obtained from ψ_σ by 'mechanical', consistent replacement of the introduced by σ constants with the original variables (if any such constants appear in ψ_σ). Hence ψ_σ is a substitution instance of ψ. The same applies to ψ'. Hence $\phi \models \psi$, and for some substitution θ, $\psi\theta \subseteq \phi$. □

Finally, the following corollary concerning completeness of bd-resolution can be stated.

Corollary 1. *Let Ψ be any normal formula. Assume that Ψ is tautology (true under any interpretation). Then there exists a bd-derivation of an empty formula $\top$ (always true) from Ψ.*

Proof. To see that the above theorem holds it is enough to put $\phi = \top$ in (5.7); thus we have that if $\top \models \Psi$ ($\models \Psi$), i.e. Ψ is valid, by Theorem 13 there exists a bd-derivation of an empty formula from it. □

To conclude, bd-resolution rule, augmented with factorization rule, constitute a tool for theorem proving which is:

- based on confirmation — an empty clause (always true one) is to be derived from the initial formula to be proved tautology;
- sound — any conclusion derived with resolution (and factorization) is sound in the sense that disjunction of parent minterms is a logical consequence of their resolvent;
- complete — in the sense that an empty clause $\top$ can always be deduced from a tautological set of minterms;
- complete with respect to logical consequence — in the sense that whenever a simple formula ϕ is a logical consequence of some normal formula Ψ, a simple formula ψ, such that $\phi \models \psi$ can be derived from Ψ.

The last statement is a general statement of completeness. It is equivalent of the so-called *Subsumption Theorem* for classical resolution.

Bd-resolution theorem proving is based on using the minterm form, i.e. quantifier-free first-order DNF formula. Hence it is especially convenient for systems which are or can be easily transformed into DNF.

5.6 Generalized Dual Resolution Rule

The presented in Definition 69 inference rule is in fact a *binary* dual resolution rule. It operates on two input formulae and produces a single output formula. The necessary condition for combining the input formulae is that they contain *complementary literals*, i.e. two literals forming a tautology. The existence of this tautology was necessary to prove soundness of bd-resolution.

Note however, that the binary version of dual resolution can be generalized over resolving more than two minterms, provided that the respective tautology can be found. The generalized dual resolution operating on k input formulae can be defined as follows.

Definition 71 (Generalized bd-resolution). *Let $M_1 = \phi_1 \wedge \omega_1$, $M_2 = \phi_2 \wedge \omega_2, \ldots, M_k = \phi_k \wedge \omega_k$ be some simple conjunctive formulae, and let σ denote a substitution. If the so-called* completeness condition *of the form*

$$\models \omega^1\sigma \vee \omega^2\sigma \vee \ldots \vee \omega^k\sigma,$$

holds, i.e. the formula $\omega^1\sigma \vee \omega^2\sigma \vee \ldots \vee \omega^k\sigma$ is a tautology, then, the generalized dual resolution rule has the form

$$\frac{\phi_1 \wedge \omega_1, \phi_2 \wedge \omega_2, \phi_k \wedge \omega_k}{\phi_1\sigma \wedge \phi_2\sigma \wedge \ldots \wedge \phi_k\sigma} . \tag{5.10}$$

The rule will be called a generalized backward dual resolution (generalized bd-resolution) *and the resulting formula is a generalized bd-resolvent of of M_1, $M_2, \ldots, M_k$.*

It can be observed that, as in the case of binary dual resolution, the disjunction of initial formulae is a logical consequence of the resulting generalized dual resolvent.

Note that it may be useful to apply some substitution to the initial formulae (factorization) before using generalized bd-resolution defined as above. Further, formulae $\omega^1, \omega^2, \ldots, \omega^k$ need not be literals.

A very interesting case is the one when the formulae $\phi_1, \phi_2, \ldots, \phi_k$ are identical, i.e. there is $\phi_1 = \phi_2 = \ldots = \phi_k = \phi$. In this case the generalized dual resolution rule takes the form of the following *gluing rule*.

Definition 72 (Gluing rule). *Let $M_1 = \phi \wedge \omega_1$, $M_2 = \phi \wedge \omega_2, \ldots, M_k = \phi \wedge \omega_k$ be some simple conjunctive formulae, and let σ denote a substitution. If the so-called* completeness condition *of the form*

$$\models \omega^1\sigma \vee \omega^2\sigma \vee \ldots \vee \omega^k\sigma,$$

holds, i.e. the formula $\omega^1\sigma\vee\omega^2\sigma\vee\ldots\vee\omega^k\sigma$ *is a tautology, then, the generalized dual resolution rule has the form*

$$\frac{\phi \wedge \omega_1, \phi \wedge \omega_2, \phi \wedge \omega_k}{\phi\sigma}\,. \tag{5.11}$$

This rule will be called a gluing rule *and the resulting formula will be called a* gluing dual resolvent.

Note that, in case of the gluing rule, not only the disjunction of initial formulae is a logical consequence of the resulting formula, but if σ is an empty substitution or σ does not influence ϕ, then also the resulting dual resolvent is a logical consequence of the input minterms. In fact, in such a case the rule preserves *logical equivalence* between the input and output formulae. Hence, the rule can be used for simplifying (reducing the size of) formulae. In further chapters we shall see the application of this rule for reduction of rule-based systems.

A further, useful generalization may be achieved by weakening the completeness condition ($\models \omega^1\sigma \vee \omega^2\sigma \vee \ldots \vee \omega^k\sigma$), which can take the form

$$\omega \equiv \omega^1\sigma \vee \omega^2\sigma \vee \ldots \vee \omega^k\sigma\ ,$$

i.e. there must exist a formula ω logically equivalent to the disjunction $\omega^1\sigma \vee \omega^2\sigma \vee \ldots \vee \omega^k\sigma$ and it can be used for replacing it. In such case the gluing rule takes the following, weaker form

$$\frac{\phi \wedge \omega_1, \phi \wedge \omega_2, \phi \wedge \omega_k}{\phi\sigma \wedge \omega\sigma}\,. \tag{5.12}$$

As before, if σ is an empty substitution or σ does not influence ϕ, then logical equivalence of input formulae disjunction and the resulting dual resolvent is preserved.

Finally, note that both of the above rules can be defined to operate under a specific interpretation (partial interpretation) referring to a certain specific world under consideration. In this way operational forms of the dual resolution principle for technical applications can be produced.

In order to provide intuitions on how the generalized dual resolution can be applied in transformation of rule-based systems, consider the following simple example of rule reduction. Consider the two following rules:

$$r_1\colon [\mathit{color} = \mathit{white}] \wedge [\mathit{shape} = \mathit{circle}] \longrightarrow [\mathit{class} = \mathit{wheel}]$$

and

$$r_2\colon [\mathit{color} = \mathit{black}] \wedge [\mathit{shape} = \mathit{circle}] \longrightarrow [\mathit{class} = \mathit{wheel}].$$

Here $\phi = [\mathit{shape} = \mathit{circle}]$, $\omega_1 = [\mathit{color} = \mathit{white}]$, $\omega_2 = [\mathit{color} = \mathit{black}]$, and, finally, $\omega = [\mathit{color} \in \{\mathit{white}, \mathit{black}\}]$. Obviously, these rules can be replaced with logically equivalent rule r of the form

$$r\colon [\mathit{color} \in \{\mathit{white}, \mathit{black}\}] \wedge [\mathit{shape} = \mathit{circle}] \longrightarrow [\mathit{class} = \mathit{wheel}].$$

Further, if there are only two colors, i.e. black and white, then the rule can be simplified to $r\colon [\mathit{shape} = \mathit{circle}] \longrightarrow [\mathit{class} = \mathit{wheel}]$, since under the assumed interpretation formula ω is always true.

The application of dual resolution for reduction of rule-based systems is discussed in Chap. 15. The schemes of possible reductions based on the gluing rule are given there. The application of dual resolution to verification of completeness is presented in Chap. 16.

Part II

Principles of Rule-Based Systems

6

Basic Structure of Rule-Based Systems

Rules are omnipresent in our life, in science and technology. We have to behave according to general rules of *good behavior* and specific rules applied in certain situation, community or organization. Rules in science express usually general laws, as in physics, for example, or allow to control technological processes in industry. There are systems described with sets of formal and informal rules[1], some of them taking precise, written form, and some of them encoded in our minds only.

Some most typical examples of rules and systems of rules are concerned with well-defined domains, they are shaped historically, and their current written form is the result of some long-lasting process.

Here are some examples:

- Law — it is divided into several domains, each of them has a certain specific set of rules; the rules are codified with the use of a bit specific language of law, enumerated, and put together so that they form the code for the domain[2].
- Traffic Code — a practical example of rules introduced to allow safe traffic of cars and pedestrians.
- Army Code — army is an example of a very formal organization; every situation and possible action are described with specific rules controlling the individual and group behavior.
- Technical equipment — always behave according to predefined rules, specifying in fact the control algorithm; for example, a vacuum cleaner behaves in a simple way (it works when switched on and does not work, when switched of), a refrigerator is a bit more complex (it also performs stabilization of required temperature through switching on automatically when

[1] A beautiful example of introducing codification into such informal domain as *personal contacts* is the famous Bodziewicz Honorary Code.

[2] A good example is the *Napoleon Code* which provided background for European Law.

A. Ligêza: *Logical Foundations for Rule-Based Systems*, Studies in Computational Intelligence (SCI) **11**, 91–96 (2006)
www.springerlink.com

the temperature inside is too high), while a lift behaves according to a certain set of rules and the current behavior depends on the status of control buttons and perhaps weight and door detector.
- Physics — a beautiful example of a number of laws existing and acting without our will, and such that we have no possibility of changing them (they are just, since they apply equally to anyone), such as the law of gravity, for example.
- Economy — there are also many rules, but usually of stochastic nature, working only in mass process.

Rules are perhaps the most universal paradigm for defining law, describing (dynamic) behavior, specifying control and decision algorithms.

6.1 Basic Concepts in Rule-Based Systems

Rule-Based Systems (RBS) provide a powerful tool for knowledge specification and development of practical applications. However, although the technology of RBS becomes more and more widely applied in practice, solutions based on the use of rule-based systems are still not well-accepted by some industrial engineers. This is so mainly due to their relationship to first-order logic and sometimes complex rule patterns and inference mechanisms. Further, the 'correct' use of them requires much intuition and domain experience, and knowledge acquisition still constitutes a bottleneck for many potential applications.

A serious problem concerning RBS is a consequence of the fact that a complete analysis of properties remains still a problem, especially one supporting the design stage rather than the final verification. This is particularly visible in case of more powerful knowledge representation languages, such as ones incorporating the full first order logic formalism.

Software systems for development of RBS are seldom equipped with tools supporting design of the knowledge-base; for some exceptions see [1, 4]. A recent, solution is proposed in [141]. A complete, new solution and a tool under the name MIRELLA has been presented in [92].

The basic form of any rule is as follows

$$rule\colon \langle preconditions \rangle \longrightarrow \langle conclusions \rangle \tag{6.1}$$

where $\langle preconditions \rangle$ is a formula defining when the rule can be applied, and $\langle conclusions \rangle$ is the definition of the effect of applying the rule; it can be logical formula, decision or action. The $\langle preconditions \rangle$ is also referred to as the *Left Hand Side* of the rule, so one writes

$$LHS(rule) = \langle preconditions \rangle,$$

and the $\langle conclusions \rangle$ is also referred to as the *Right Hand Side* of the rules, so one writes also

$$RHS(rule) = \langle conclusions \rangle.$$

The most popular form of a rule has some number of atomic preconditions (say n) and a single conclusion; such a rule can be expressed as

$$p_1 \; AND \; p_2 \; AND \; \ldots \; AND \; p_n \longrightarrow h \qquad (6.2)$$

In the above rule $p_1, p_2, \ldots, p_n$ are atomic formulae of some accepted language (e.g. propositional logic, attributive logic, first order logic) and h is the conclusion, action or decision. There are also more complex forms of rules (to be discussed later).

Depending on the accepted language, the rule can have different expressive power. A very popular solution is based on the use of various forms of attribute-based languages. This makes the notation similar to the one used in Relational Database Systems (RDBS). However, contrary to Rule-Based Systems, Relational Data Base Systems offer relatively simple, but matured data manipulation technology, employing widely accepted, intuitive knowledge representation in tabular form. It seems advantageous to make use of elements of this technology for simplifying certain operations concerning RBS.

Note, for example, that from practical point of view any row of a RDBS table can be considered as a rule, provided that at least one attribute has been selected as an output (and there is a so-called *functional dependency* allowing for determination of the value of this attribute on the base of some other attributes). Thus, it seems that merging elements of RBS and RDBS technologies can constitute an interesting research area of potential practical importance.

An idea suggested in this book is to investigate RBS by means of RDBS-like tabular knowledge representation and algebraic rather than logical tools. A relatively simple approach derived from first-order logic, but incorporated into a RDBS-like framework, is also used in this work; details will be discussed in Chap. 8. Further, a hierarchical structure (or even a network-like one) of the RBS can be assumed. For an example, consider a two-level system structure presented below.

The organization of the system is as follows:

- There is an upper level consisting of several *contexts* of work, defined with formulae $C_1, C_2, \ldots, C_c$; selection of the context depends on current working conditions and the goal. Selection of the current context can be performed by a meta-level decision mechanism, while switching among contexts can be a side effect of lower level rule application.
- For any context C_i there is a simple, uniform (i.e. using the same scheme of attributes), tabular RBS, similar to a decision table, with knowledge representation based on attributes. When the context is selected, the system identifies and applies a single rule; then the cycle is repeated.

The system is organized in a hierarchical way; in the simplest case above one has just two-level specification. The upper level provides context selection

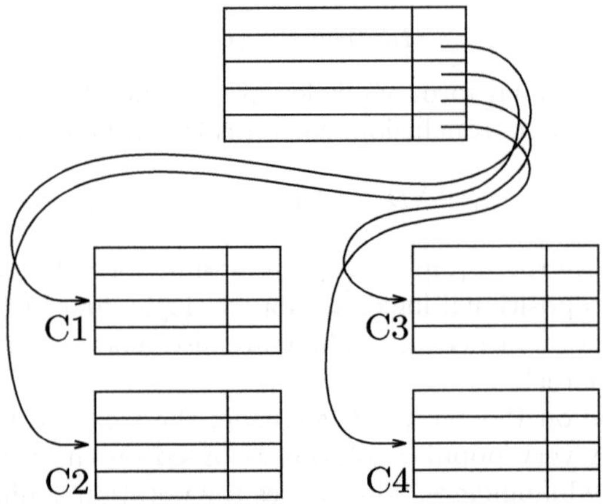

Fig. 6.1. Graphical presentation of the idea of a hierarchical, two-level tabular rule-based system

mechanism, while the lower level is responsible for rule selection and application. A graphical presentation of the idea of hierarchical system is presented in Fig. 6.1.

In general, one can consider any more complex scheme, including several levels and numerous tables connected according to various patterns. Note that such a multi-tabular system can be (at least potentially) reduced to one big table, similarly as in the case of RDB systems. However, it would not be a reasonable approach, both from the knowledge representation and the analysis point of view — most of the attributes would be useless in most of the rules. Instead, it seems much more appropriate to analyze any particular uniform tabular system within the appropriate context C_i. Thus, the discussion presented in this work will often be restricted to the level of a single tabular component; note however, that the meta-knowledge concerning context switching can also be specified with use of appropriate tables.

A single-table system is assumed to be a forward-chaining one, operating according to the following scheme: for given input situation, an applicable rule is searched for, and if found, the rule is fired. The environment of the system may be changed by the system itself, or changes may be due to dynamic environment, as in the case of rule-based control systems [53, 56]. For the idea see Fig. 6.2.

In this work we shall also address the issues emerging during logical verification of theoretical properties especially in case of such single-layer rule-based reactive systems; an intuitive, geometric interpretation of the work of the system is presented in Fig. 6.3.

RBS defined and working as above constitute a class applicable in a wide spectrum of control and decision support tasks [48, 53, 81]. The right-hand side of any rule is normally a control action (sometimes also an assertion to

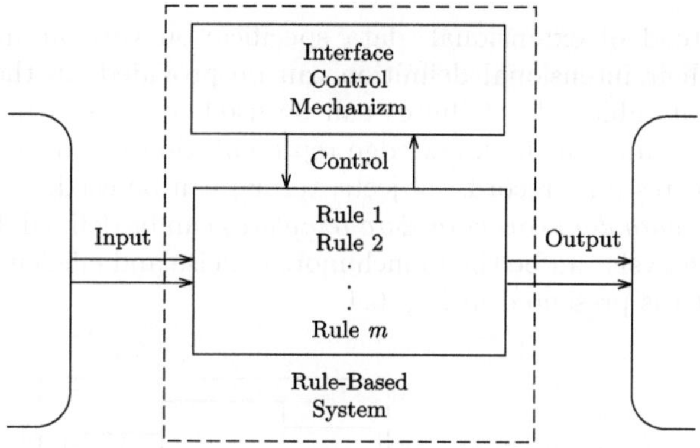

Fig. 6.2. A general scheme of single-level rule-based system

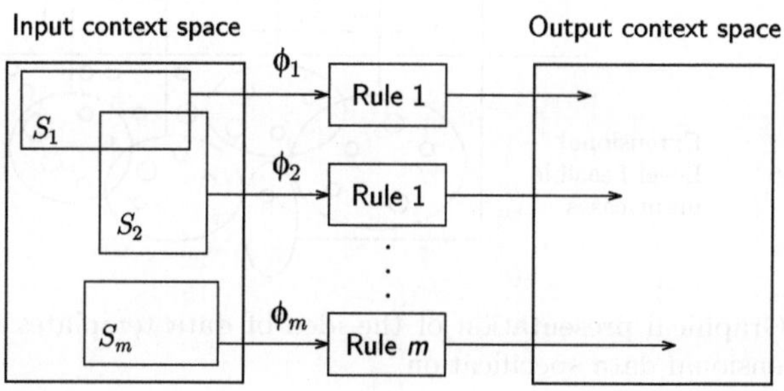

Fig. 6.3. An abstract geometric presentation of the working scheme of a rule-based control system

or deletion from the fact base) or decision; no direct chaining among the rules takes place. Their standard working cycle proceeds as follows: the current state of the input environment is observed, then a single rule matching the input pattern is selected, and, finally, the selected rule is executed. The whole cycle is repeated in a closed loop.

Systems as above form a class of simple forward reasoning expert systems with no chaining among rules (application of a single rule results in conclusions/actions). Such systems were discussed in [53, 57, 81]; many example applications are given in [48]. An important area of applications include intelligent control systems with control knowledge specified as a RBS. They can also constitute the lowest part of more complex, hierarchically structured systems where they constitute the core inference tool for any context determined at higher level. For intuition, a system as the one presented in Fig. 6.3 is complete, if the rectangles referring to rule precondition formulae cover the input space.

The tabular systems discussed further in this book can also be used as extended RDB paradigm for unconditional knowledge specification. In such a way instead of extensional, data specification with atomic values of attributes, their intensional definition can be provided. In the basic case, set and interval values of attributes can be used to cover a number of specific cases. Depending on the knowledge representation language, also more complex structures (e.g. records, objects, terms) can be used. In such a way *data patterns* or *data data covers* or *data templates* can be defined. Both representation and analysis can be then much more concise and efficient. An illustration for this idea is presented in Fig. 6.4.

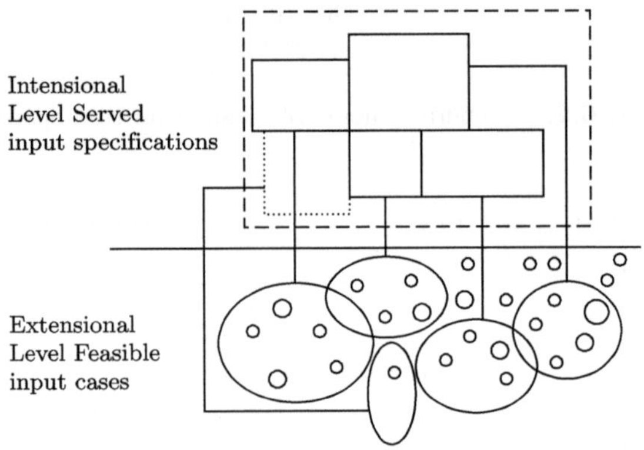

Fig. 6.4. Graphical presentation of the idea of data templates representation replacing extensional data specification

Note that there are many common points in the analysis of such intensional, unconditional data representation and the verification of tabular RBS properties. For example, checking for determinism of a rule based systems requires verification if their precondition formulae are defining separate sets of states; the same check can be performed to verify if certain data templates describe disjoint sets of data. Analogous situation occurs in checks of completeness, etc. This allows to present and discuss the problems of analysis simultaneously for data templates (extended database paradigm) end tabular systems using a common model for data and knowledge.

7

Rule-Based Systems in Propositional Logic

Although *Propositional Calculus* is possibly the simplest logical system, both with respect to syntax as well as semantics, it can serve as a practically useful language for encoding rule-based systems. Obviously, since no individual variables nor terms are allowed, knowledge representation capability is drastically limited. Simultaneously, due to very simple knowledge representation, knowledge processing requires only very simple mechanisms[1], and as such — reasoning with such rules can be relatively fast. Further, both analysis and design of such systems are relatively simple. But what is most important, majority of the mechanisms incorporated in rule-based systems can be presented and discussed with this simple model.

In this chapter we try to take the advantage of very simple form of propositional rule-based systems so as to introduce basic ideas incorporated in any more advanced logical system with rule component. In fact, propositional rule-based systems can also serve as a basic model for rule-based systems and most of discussions around them.

Propositional rule-based systems can take various visual forms incorporating some structural representation. In this chapter, apart from pure rule-based systems in the form of a set of rules (an ordered set of rules) we present some other forms such as: decision lists, decision trees, binary decision diagrams, and tabular systems.

7.1 Notation for Propositional Rule-Based Systems

Basically, the alphabet and symbols for defining propositional sets of rules are the ones of *Propositional Logic*, as introduced in Chapter 1. Recall that apart propositional logic formulae symbols, two special symbols for denoting a formula which is always true, say $\top$, and a formula which is always false, say

[1] For example, since no variables are in use, no unification mechanism at the level of terms is necessary — two propositional symbols are either identical or different.

A. Ligęza: *Logical Foundations for Rule-Based Systems*, Studies in Computational Intelligence (SCI) 11, 97–127 (2006)
www.springerlink.com

$\bot$, will be used. The atomic formulae will be denoted mostly with propositional symbols such as p, q, r, etc. and we shall assume that a set P of such symbols is defined as $P = \{p, q, r, \ldots, p_1, q_1, r_1, \ldots, p_2, q_2, r_2, \ldots\}$.

As introduced in Chap. 1, in order to assign some precise meaning to propositional symbol p, for example 'It is cold', the following notation can be used

$$p \stackrel{\text{def}}{=} \text{'It is cold'} .$$

Recall also that any propositional variable can be assigned a unique meaning only in the considered application. Further, two propositional variables may be *independent* or *dependent* on each other. They are independent if the assigned interpretations (truth-value assignments) are independent; in such a case the variables can take logical values independently on each other. They are dependent if the interpretation of one of them known to be true (false) implies that the other interpretation is known. For example, if p is assigned the meaning as above, and q is assigned meaning as

$$q \stackrel{\text{def}}{=} \text{'It is dark'} ,$$

the truth-assignment to p and q are independent — in real world it can be simultaneously cold and dark, only cold, only dark, or none of these possibilities. However, if r is assigned meaning as

$$r \stackrel{\text{def}}{=} \text{'The temperature outside is minus 25 degrees Centigrade'} ,$$

then p and q are no longer independent formulae. Most of us agree that r logically implies p, i.e.

$$r \models p .$$

In the following part it will be assumed that the considered atomic propositional formulae are independent, unless stated explicitly that some kind of dependency takes place. In case logical implication holds, in most cases it will be stated with an explicit rule of the form

$$r \longrightarrow p ,$$

or if for some reasons no explicit rule is present, the logical entailment symbol will be used to state that $r \models p$.

7.2 Basic Propositional Rules

The most basic logical form of propositional rules is as follows

$$rule\colon p_1 \wedge p_2 \wedge \ldots \wedge p_n \longrightarrow h . \tag{7.1}$$

Recall that such a form of a rule, defined previously by (1.10) is logically equivalent to a Horn clause given by (1.9) provided that all the literals are positive.

The $LHS(rule) = p_1 \wedge p_2 \wedge \ldots \wedge p_n$ is a conjunction of propositional literals; each p_i may be a positive atom or a negated one. The $LHS(rule)$ part of the rule defines its *preconditions*, i.e. conditions which must simultaneously hold to fire the rule. The $RHS(rule) = h$ defines the conclusion of the rule which is a single positive or negative literal.

A more complex rule may contain conclusion part composed of several propositions. In such a case the rules are defined as follows

$$rule\colon p_1 \wedge p_2 \wedge \ldots \wedge p_n \longrightarrow h_1 \wedge h_2 \wedge \ldots h_k\,. \tag{7.2}$$

As before, the $LHS(rule) = p_1 \wedge p_2 \wedge \ldots \wedge p_n$ is a conjunction of propositional literals. The $RHS(rule) = h_1 \wedge h_2 \wedge \ldots h_k$ defines the conclusion of the rule which is now a conjunction of positive or negative literals.

If the precondition part of a rule is composed of both positive and negative literals, then it can be decomposed as follows

$$LHS(rule) = LHS^{+}(rule) \wedge LHS^{-}(rule)\,,$$

where $LHS^{+}(rule)$ denotes the positive literals and $LHS^{-}(rule)$ — negative ones. In certain systems only positive specifications of preconditions are allowed ($LHS(rule) = LHS^{+}(rule)$). This is a very typical simplification; negative literals, if necessary, are replaced with positive equivalents. Note however, that in this approach at least some propositional symbols are no longer independent.

A further simplification frequently met in practice is that the conclusion part consists of a single literal. In fact, the rule specified with (7.2) can be equivalently transferred to a set (logically: conjunction) of rules having single conclusions (from logical point of view these are Horn clauses) of the following form:

$$\begin{aligned}
&rule_1\colon p_1 \wedge p_2 \wedge \ldots \wedge p_n \longrightarrow h_1\,,\\
&rule_2\colon p_1 \wedge p_2 \wedge \ldots \wedge p_n \longrightarrow h_2\,,\\
&\vdots\\
&rule_k\colon p_1 \wedge p_2 \wedge \ldots \wedge p_n \longrightarrow h_k\,.
\end{aligned} \tag{7.3}$$

In fact, considering $\longrightarrow$ as the equivalent of $\Rightarrow$ (implication) one can write (7.2) in the following form

$$\neg(p_1 \wedge p_2 \wedge \ldots \wedge p_n) \vee (h_1 \wedge h_2 \wedge \ldots \wedge h_k).$$

By subsequent application of the distributivity law one gets:

$$\begin{aligned}
&(\neg(p_1 \wedge p_2 \wedge \ldots \wedge p_n) \vee h_1)\wedge\\
&(\neg(p_1 \wedge p_2 \wedge \ldots \wedge p_n) \vee h_2)\wedge\\
&\vdots\\
&(\neg(p_1 \wedge p_2 \wedge \ldots \wedge p_n) \vee h_k).
\end{aligned}$$

Finally, we can go back to the rule form:

$$
\begin{aligned}
&((p_1 \wedge p_2 \wedge \ldots \wedge p_n) \longrightarrow h_1)\wedge \\
&((p_1 \wedge p_2 \wedge \ldots \wedge p_n) \longrightarrow h_2)\wedge \\
&\vdots \\
&((p_1 \wedge p_2 \wedge \ldots \wedge p_n) \longrightarrow h_k)\ ,
\end{aligned}
$$

which is just another presentation of (7.3). The reverse transformation is also straightforward — it is enough to inverse the procedure. Note that although logical equivalence is kept, the operation of systems with rules of multi-literal conclusions and their single-literal equivalent may not be the same. This depends on the rule inference control mechanism. In case of rule defined by (7.2) all the conclusions $h_1 \wedge h_2 \wedge \ldots h_k$ are deduced at once, simultaneously. In case of the set of rules given by (7.3) it may be the case that only one rule (or a subset of them) will be fired; in such a case not all the conclusions are stated valid. Moreover, the cumulated result of executing the rules may depend on the order of execution.

7.3 Propositional Rules with Complex Precondition Formulae

In certain practical applications it may happen that the preconditions of rules are defined as arbitrarily complex formulae of propositional logic. In such a case, instead of the form given by (7.1) any rule can be represented as

$$rule\colon \Phi \longrightarrow h\ . \tag{7.4}$$

In this case checking satisfaction of the precondition formula Φ may become a bit more complex. Thus, if such a rule appears it is customary to simplify it by replacing with simple rules having the form defined by (7.1).

The first step is accomplished by transforming formula Φ to its Disjunctive Normal Form (DNF), as defined by (Def. 20). The steps of the transformation procedure are defined in Sect. 1.6.5. Now the rule looks as follows

$$rule\colon \phi_1 \vee \phi_2 \vee \ldots \vee \phi_n \longrightarrow h\ .$$

Considering it as a logical implication, one can rewrite it as

$$rule\colon \neg(\phi_1 \vee \phi_2 \vee \ldots \vee \phi_n) \vee h\ ,$$

and after applying the De Morgan's law we obtain

$$rule\colon (\neg\phi_1 \wedge \neg\phi_2 \wedge \ldots \wedge \neg\phi_n) \vee h\ .$$

Next, by applying the distributivity law it is possible to obtain the following form:

$$rule\colon (\neg\phi_1 \vee h) \wedge (\neg\phi_2 \vee h) \wedge \ldots \wedge (\neg\phi_n) \vee h)\ ,$$

which finally is equivalent to

$$rule\colon (\phi_1 \Rightarrow h) \wedge (\phi_2 \Rightarrow h) \wedge \ldots \wedge (\phi_n \Rightarrow h) .$$

Finally, the rule can be replaced with n simple rules of the form:

$$\begin{aligned} rule_1 &\colon \phi_1 \longrightarrow h , \\ rule_2 &\colon \phi_2 \longrightarrow h , \\ &\vdots \\ rule_n &\colon \phi_n \longrightarrow h , \end{aligned} \tag{7.5}$$

where each ϕ_i is a simple conjunction of literals. The final set is of the form given by (7.3).

7.4 Activation of Rules

Activation of rules depends on two main factors which are:

1) whether the preconditions of the rule are satisfied,
2) whether the rule is selected by the inference engine.

Rules are selected and examined by the inference engine according to some specific, predefined algorithm; this may be for example linear scanning, linear scanning in a closed loop, selection according to some preferences, parallel execution, etc. A single rule, when examined is tested if the preconditions of it are satisfied. If so the rule is executed (fired) or selected for further examination in case several rules have satisfied preconditions at the same time. Such rules having simultaneously satisfied preconditions form the so-called *conflict set*, since in most systems only one rule may be fired at a time.

Let ϕ define the current state of the considered system; a state is defined as a specification of all the facts which are true (positive) and false (negative) at a precise instant of time. Hence the state-defining formula can be decomposed into a conjunction of positive literals and another one composed of negative literals; so we have

$$\phi = \phi^+ \wedge \phi^- ,$$

where ϕ^+ and ϕ^- are the respective components of true and false atoms, respectively.

Now the condition defining the possibility to fire a selected rule is as follows:

$$\phi \models LHS(rule) \tag{7.6}$$

which, according to the definition of satisfaction, can be split into the two following simpler conditions:

$$\phi^+ \models LHS^+(rule) , \tag{7.7}$$

$$\phi^- \models LHS^-(rule) . \tag{7.8}$$

Note that in practical solutions the formulae defining state are specified as two separated lists (sets) of positive and negative atoms, $[\phi^+]$ and $[\phi^-]$, respectively, where $[\phi]$ denotes the set of literals occurring in formula ϕ. Hence, the above logical conditions of rule satisfaction can be replaced with simple algebraic conditions of the form:

$$LHS^+(rule) \subseteq [\phi^+] \,, \tag{7.9}$$

$$LHS^-(rule) \subseteq [\phi^-] \,. \tag{7.10}$$

Taking into account that an atomic formula can be either true or false in current state, one can also specify simple algebraic conditions (a kind of a test) for excluding the application of a rule; they take the obvious form as follows:

$$LHS^+(rule) \cap [\phi^-] \neq \emptyset \,, \tag{7.11}$$

$$LHS^-(rule) \cap [\phi^+] \neq \emptyset \,. \tag{7.12}$$

The above criteria (7.11) and (7.12) can be used for relatively fast elimination of candidate rules rather than checking for their satisfaction — in such a case it is enough to find *a single* atom belonging to the opposite list rather than checking if *all* of them are specified in the appropriate set.

7.5 Deducibility and Transitive Closure of Fact Knowledge Base

Consider a set of propositional rules, say $R = \{r_1, r_2, \ldots, r_m\}$. For simplicity, assume that any rule $r_i \in R$ is of the simplest form defined by (7.1). Let there be also given a *Fact Base* (*FB*), $FB = \{q_1, q_2, \ldots, q_q\}$ denoting a given set of facts defined with literals $q_1, q_2, \ldots, q_q$; logically, it is assumed that the conjunction of all the facts given by *FB* is true, i.e.

$$\phi = q_1 \wedge q_2 \wedge \ldots \wedge q_q$$

holds.

In classical, forward-chaining systems the rules are applied by checking if their preconditions are satisfied and firing. Whenever a rule is fired, its conclusion is added to the current state. The satisfaction of rule preconditions is defined by (7.6), and the practical tests are given thereafter.

We shall say that fact h is *deducible* from fact base *FB* by rule r_i iff r_i is of the form

$$r_i\colon p_1 \wedge p_2 \wedge \ldots \wedge p_n \longrightarrow h$$

and the rule is satisfied in the state defined by *FB*, i.e. there is $LHS(r_i) \subseteq FB$. Obviously, if the rule is considered as logical implication, this scheme of rule application is analogous to the well-known *Modus Ponens* rule. More precisely, the operation of applying a rule to fact base can be specified as follows

$$\frac{q_1 \wedge q_2 \wedge \ldots \wedge q_q, p_1 \wedge p_2 \wedge \ldots \wedge p_n \longrightarrow h}{q_1 \wedge q_2 \wedge \ldots \wedge q_q, h} \tag{7.13}$$

provided that $\{p_1, p_2, \ldots, p_n\} \subseteq \{q_1, q_2, \ldots, q_q\}$. For practical reasons, the last check is usually performed in the form specified by (7.9) and (7.10).

If a rule r is applicable to fact base FB^i, as the result of application of this rule a new fact base FB^{i+1} is obtained; we shall write

$$r\colon FB^i \longrightarrow FB^{i+1} \tag{7.14}$$

or simply $r(FB^i) = FB^{i+1}$. The new fact base is defined as

$$FB^{i+1} = FB^i \cup \{h\} \tag{7.15}$$

or, if the rule has more conclusions, i.e. it is of the form (7.2), all of them are added to the fact base

$$FB^{i+1} = FB^i \cup \{h_1, h_2, \ldots, h_k\}\,. \tag{7.16}$$

The inference process in rule based systems consists of sequential selection, matching, and application of rules to some initial fact base. Let us define the mechanism of simple, flat inference engine, operating in one step through applying all rules in turn to a given fact base; the conclusions which are generated are added after applying the last rule to the initial base.

Definition 73. *Let R be a finite set of propositional rules, $R = \{r_1, r_2, \ldots, r_m\}$ and let FB^i and FB^{i+1} be two fact bases. We shall say that fact base FB^{i+1} is a result of application of the set of rules R to FB^i, and so we shall write*

$$FB^{i+1} = R(FB^i) \tag{7.17}$$

if

$$FB^{i+1} = FB^i \cup \{h\colon \textit{for some } r \in R, FB^i \models LHS(r) \textit{ and } h \in RHS(r)\}. \tag{7.18}$$

So FB^{i+1} contains all the conclusions which can be produced from FB^i by applying the rules of R to FB^i, including the facts of FB^i.

In classical systems new facts are added to the fact base when they are deduced. This enlarges the fact base so that perhaps new rules can be applied.

Definition 74. *Let R be a finite set of propositional rules and let FB^0 and FB^k be two fact bases. Fact base FB^k is deducible from fact base FB^0 with rules of R if and only if there exists a finite sequence of rules $r^1, r^2, \ldots, r^k \in R$ together with a finite sequence of fact bases $FB^1, FB^2, \ldots, FB^{k-1}$, such that:*

$$\begin{aligned} r^1:&\ FB^0 &\longrightarrow FB^1, \\ r^2:&\ FB^1 &\longrightarrow FB^2, \\ &\vdots \\ r^k:&\ FB^{k-1} &\longrightarrow FB^k\ . \end{aligned} \tag{7.19}$$

Note that the process of generating new fact bases is monotonic, i.e. every newly generated fact base must contain all the facts covered by any former fact base. Obviously, if a sequence of fact bases is generated according to (7.19), then there is

$$FB^0 \subseteq FB^1 \subseteq \ldots \subseteq FB^k\ .$$

Finally, since the number of rules is finite, and so is the number of propositional symbols used in a specific example, the process of generating new fact bases is expected to stop at some point. We define the fixed-point for this operation as follows.

Definition 75. *Consider a finite set of propositional rules R which are applied in turn to an initial fact base FB^0. In this way a sequence of fact bases $FB^0 \subseteq FB^1 \subseteq \ldots \subseteq FB^k$ is generated. The fact base being the fixed-point of this operation is defined as*

$$R(FB^k) = FB^k \tag{7.20}$$

and it is such a set FB^k that for any rule $r \in R$, $r(FB^k) = FB^k$. The fixed point FB^k will be denoted as $R^*(FB^0)$ or FB^*, for short.

For intuition, the fixed point FB^* is the maximal set of facts which can be deduced from the initial fact base with the rules of R — no more fact can be deduced. The set is obtained by a level-saturation method. Further, for monotonic systems this set is defined in a unique way.

Theorem 14. *Consider a finite set of propositional rules R which are applied in turn to an initial fact base FB^0 so that the fixed point $R^*(FB^0) = FB^k$ is reached for some k. The set $R^*(FB^0)$ is defined in a unique way.*

Proof. The proof is by induction with respect to the iteration number i. Consider a fact base FB^i obtained as the result of iteration i; the basic goal is to show that FB^{i+1} obtained at the next iteration is defined in a unique way.

Recall that FB^{i+1} is generated according equation (7.18). Note that, since all the rules operate on the same fact base FB^i during one iteration, the set of generated conclusions is independent on the order of rule application. Hence FB^{i+1} is defined in a unique way.

By finite induction with $i = 1, 2, \ldots, k$ we conclude that also FB^k must be defined in a unique way. □

Note that the inference process and reaching the medium fact bases FB^i as well as the final fixed point base FB^* can usually be reached faster, in a more efficient way than by a procedure operating level-by-level according to (7.18). The idea is that after selection and application of a rule $r \in R$ to some fact base FB the concluded facts are added immediately to the base, so that $r(FB) = FB'$, where $FB \subseteq FB'$. Obviously, the new facts in FB' allow for application of the same rules as applicable to FB, but perhaps some more rules can also be applied. In this way, after one iteration starting at some fact base FB^i the resulting next stage base $(FB^{i+1})'$ can be bigger than the one obtained according to 7.18. However, although depending on the order of rule application one can obtain different medium-stage fact bases, the fixed point FB^* remains unique, and the same as previously.

We have the following Corollary of Theorem 14.

Corollary 2. *The fixed point FB^* is unique, independent from the rule order and method of adding newly generated facts, provided that all the rules are applied in a systematic way (i.e. all the rules are applied during each iteration).*

Proof. Let the inference process procedure produce the following sequence of fact bases: $FB^0, (FB^1)', (FB^2)', \ldots, (FB^k)'$. As shown above, one can always generate a corresponding sequence of fact bases $FB^0, FB^1, FB^2, \ldots, FB^k$ according to 7.18. Obviously, there is $FB^i \subseteq (FB^i)'$. Let the sequences be generated so long that $FB^k = FB^*$, i.e. the fixed point is reached. At this point it must be $(FB^k)' = FB^k$, since $FB^k \subseteq (FB^k)'$, but simultaneously no bigger set than FB^k can be generated — according to Definition 75, there is $r(FB^k) = FB^k$ for any rule $r \in R$. □

The above properties of reaching the fixed point fact base which is defined in a unique way is valid thanks to *monotonicity* of classical logical systems and the deduction rules; in our case defined by (7.15) and (7.16). In the next sections we discuss non-monotonic systems, where these properties do not hold any longer.

7.6 Various Forms of Propositional Rule-Based Systems

In this section we shall present various visual forms of propositional rule-based systems. Such forms incorporate elements of a structure which plays a double role:

1) it provides an intuitive visualization of the system, usually oriented towards some specific application (such as decision making);
2) it forces specific interpretation of the rules, especially with respect to (i) the order of rules and (ii) the order of preconditions.

Five most typical representations will be discussed below. These are:

1) Decision Tables,
2) Decision Lists,
3) Decision Rules with control statements,

4) Decision Trees,
5) Binary Decision Diagrams (BDD).

From logical point of view, all of the above forms model a set of rules of the form given by (7.1) limited — for simplicity of the discussion — to a single conclusion propositional symbol h, which may take value *true* or *false* (1 or 0).

Note that the set of rules may, in general, incorporate rules with different preconditions, and of different length of the *LHS* part. The discussion below applies directly to such rules, however, for simplicity and elegance of the presentation, it will be assumed that all the rules use the same propositional symbols, ordered according to the same scheme, and different only with respect to using the symbol of negation before the propositional symbol or not. Hence, all the rules are assumed to be of the following scheme:

$$\begin{aligned} &rule_1\colon \#p_1 \wedge \#p_2 \wedge \ldots \wedge \#p_n \longrightarrow \#h_1,\\ &rule_2\colon \#p_1 \wedge \#p_2 \wedge \ldots \wedge \#p_n \longrightarrow \#h_2,\\ &\vdots\\ &rule_m\colon \#p_1 \wedge \#p_2 \wedge \ldots \wedge \#p_n \longrightarrow \#h_m, \end{aligned} \tag{7.21}$$

where # means *nothing* (so we have unnegated propositional symbol) or # means $\neg$ (and so the following propositional symbol is negated).

A set of rules given by (7.21) with preconditions which differ only with respect to using or not the negation sign will be referred to as *canonical* set of rules, and if all the possible combinations of using the negation sign are present (the disjunction of all the preconditions, i.e. $LHS(rule_1) \vee LHS(rule_2) \vee \ldots LHS(rule_k)$, is tautology) it will be referred to as *complete* or *full canonical* set of rules.

Obviously, for any set of rules which is not in canonical form, one can always construct a logically equivalent system in canonical form. The transformation is basically the same as transformation of any logical formula in CNF into a formula in maximal CNF (with maximal minterms). Each of the rules can be transformed separately into an appropriate set of rules according to the following principles:

1. For any rule with precondition formula ϕ identify all the propositional symbols $q_1, q_2, \ldots, q_j$ which do not occur in its precondition formula (either negated or unnegated).
2. For any such symbol build a disjunction of the form $(q_i \vee \neg q_i)$, $i = 1, 2, \ldots, j$, which is obviously tautology.
3. Put all the tautological disjunctions into the precondition formula to build an equivalent formula of the form $\phi \wedge (q_1 \vee \neg q_1) \wedge (q_2 \vee \neg q_2) \wedge \ldots \wedge (q_j \vee \neg q_j)$.
4. Using the distributivity law transform the precondition formula to an appropriate max DNFform, and split the rule to 2^j equivalent rules, as explained in Sect. (7.3).

5. Finally, order the literals in preconditions of the set of rules in a unique way.

Let us explain the procedure with a simple but illustrative example. Consider two rules specified as below:

$$\begin{aligned} \mathit{rule}_1 &: p \longrightarrow h, \\ \mathit{rule}_2 &: q \longrightarrow h. \end{aligned} \tag{7.22}$$

The rules have completely independent preconditions. In order to transform them to canonical form we complete the preconditions according to the following scheme:

$$\begin{aligned} \mathit{rule}_1 &: p \wedge (q \vee \neg q) \longrightarrow h, \\ \mathit{rule}_2 &: q \wedge (p \vee \neg p) \longrightarrow h. \end{aligned} \tag{7.23}$$

After applying the distributivity laws we obtain:

$$\begin{aligned} \mathit{rule}_1 &: (p \wedge q) \vee (p \wedge \neg q) \longrightarrow h, \\ \mathit{rule}_2 &: (q \wedge p) \vee (q \wedge \neg p) \longrightarrow h. \end{aligned} \tag{7.24}$$

Finally after splitting the rules to ones with simple conjunctive preconditions and ordering the literals in preconditions in a unique way we obtain:

$$\begin{aligned} \mathit{rule}_1^1 &: p \wedge q \longrightarrow h, \\ \mathit{rule}_1^2 &: p \wedge \neg q \longrightarrow h, \\ \mathit{rule}_2^1 &: p \wedge q \longrightarrow h, \\ \mathit{rule}_2^1 &: \neg p \wedge q \longrightarrow h. \end{aligned} \tag{7.25}$$

Note that among the rules in canonical form, the first one and the third one are identical; one of them can be deleted without influencing the logical equivalence. Hence, the final set of canonical rules is of the form:

$$\begin{aligned} \mathit{rule}_1^1 &: p \wedge q \longrightarrow h, \\ \mathit{rule}_1^2 &: p \wedge \neg q \longrightarrow h, \\ \mathit{rule}_2^1 &: \neg p \wedge q \longrightarrow h. \end{aligned} \tag{7.26}$$

To summarize, the following corollary can be formulated.

Corollary 3. *Let $PROP(\phi)$ denote the set of different propositional symbols occurring in formula ϕ. For any set of k rules of the form $\phi_i \longrightarrow h$ there exists an equivalent set of rules in canonical form given by (7.21) where $PROP(\phi_1) \cup PROP(\phi_2) \cup \ldots PROP(\phi_k) = \{p_1, p_2, \ldots, p_n\}$. The number of rules m in the canonical form satisfies the inequality $k \leq m \leq 2^n$.*

Note that, the above transformation does not make use of the *RHS* part of the rules. Hence, in fact any set of rules (with different conclusions) can be transformed to a canonical form with respect to preconditions.

Such a transformation, although always possible from computational point of view, may lead to an unexpectedly large number of rules, limited by 2^n, where n is the number of propositional symbols in use. Hence, it seems reasonable to keep together in one set rules operating in certain context, and thus having similar preconditions by initial specification.

7.6.1 Example

Finally, consider a simple example of canonical set of propositional rules. Let us consider the following formula[2]

$$(p \Leftrightarrow q) \wedge (r \Leftrightarrow s) \, .$$

Let us specify a set of rules for deciding if the above formula is satisfied.

Assuming we have one conclusion symbol h meaning that the formula is satisfied, a single rule can be written as

$$(p \Leftrightarrow q) \wedge (r \Leftrightarrow s) \longrightarrow h \, .$$

In case we want to have a set of rules with preconditions being simple conjunctive formulae, we have to transform the initial rule to a canonical set of rules; it is of the following form:

$$\begin{aligned} rule_0 &: \neg p \wedge \neg q \wedge \neg r \wedge \neg s \longrightarrow h, \\ rule_3 &: \neg p \wedge \neg q \wedge r \wedge s \longrightarrow h, \\ rule_{12} &: p \wedge q \wedge \neg r \wedge \neg s \longrightarrow h, \\ rule_{15} &: p \wedge q \wedge r \wedge s \longrightarrow h \, . \end{aligned} \tag{7.27}$$

The rules are enumerated in a way analogous to the construction of binary numbers and their translation to decimal numbers; if $pqrs$ is a sequence of zeros and ones satisfying precondition formula of a rule, then the rule number is calculated as $p * 2^3 + q * 2^2 + r * 2^1 + s * 2^0$.

Note that if we want to have a complete specification, we need extra 12 rules which are as follows:

[2] The example is inspired by [2]; through the next subsections we shall continue with this example on, so as to show some selected different forms of representation of rule-based systems.

$$
\begin{aligned}
&rule_1: \neg p \wedge \neg q \wedge \neg r \wedge s \longrightarrow \neg h,\\
&rule_2: \neg p \wedge \neg q \wedge r \wedge \neg s \longrightarrow \neg h,\\
&rule_4: \neg p \wedge q \wedge \neg r \wedge \neg s \longrightarrow \neg h,\\
&\vdots\\
&rule_{11}: p \wedge \neg q \wedge r \wedge s \longrightarrow \neg h,\\
&rule_{13}: p \wedge q \wedge \neg r \wedge s \longrightarrow \neg h,\\
&rule_{14}: p \wedge q \wedge r \wedge \neg s \longrightarrow \neg h.
\end{aligned}
\tag{7.28}
$$

In the above set rules with numbers 0, 3, 12 and 15 are omitted.

The above set of rules given by (7.27) and (7.28) is a complete, canonical set of rules. Although mathematically elegant, it is certainly not minimal. However, the problem of minimization will be discussed in a separate Chapter further on.

7.6.2 Binary Decision Tables

Binary decision tables incorporate the basic ideas of propositional rule-based systems in their basic form. A decision table represents in fact a set of simple propositional rules grouped together and similar with respect to used preconditions and conclusions or actions. In fact, typical decision tables make use of a canonical set of rules, as defined by (7.21).

The rules covered by a decision table are put all together and form a kind of *decision unit* designed to work in some assumed context situation.

A classical decision table is a table displaying sequences of conditions which must hold for executing specific actions or drawing specific conclusions. The sequences of conditions are displayed in a readable form, vertically in the classical decision tables [120], as parts o columns of the decision table, or horizontally, as parts of rows of the table.

A classical form of decision table [120] (Table 7.1) is the vertical one, where *condition_i* specifies the condition to be examined, and *action_i* defines the action to be executed. The values of logical conditions specified in the leftmost column are specified with v_{ij} and in the basic form of decision tables can take the following basic values:

- With respect to conditions:
  - +, T or Y if the condition must hold for certain action to be executed;
  - −, F or N if the condition cannot hold (if it holds, then the action cannot be executed);
  - _ or − if the execution of the action does not depend on the specific condition.

  With respect to conclusions:
  - X or + if the action should be executed (or the conclusion should be drawn);

Table 7.1. The form of classical decision table with vertical rules

	rule_1	*rule_2*	...	*rule_m*
condition_1	v_{11}	v_{12}	...	v_{1m}
condition_2	v_{21}	v_{22}	...	v_{2m}
$\vdots$	$\vdots$	$\vdots$		$\vdots$
condition_n	v_{n1}	v_{n2}	...	v_{nm}
action_1	w_{11}	w_{12}	...	w_{1m}
action_2	w_{21}	w_{22}	...	w_{2m}
action_1	$\vdots$	$\vdots$		$\vdots$
action_k	w_{k1}	w_{k2}	...	w_{km}

– _ or – if the action should be ignored.

The evaluation of the table proceeds as follows:

- any column of values specifying action prerequisites is traversed top-down, and the defined sequence of true and false conditions is verified;
- if the pattern is matched by the current state, the actions specified below are executed;
- next subsequent column of conditions is analyzed in a similar way.

Note that, when operating in a stable environment (most of the off-line, decision support applications), the conditions can be evaluated once for processing of all the table, and then only conditional sequences are matched against the current pattern of true and false conditions. This can save time and reduce repeated computational effort, especially in case of large applications and multiple users accessing the same data. However, in on-line, dynamic applications it is likely that for every rule the preconditions must be evaluated immediately before its application. This is especially the case if one of the formerly fired rules might influence the conditions directly through its actions (or as a side-effect).

Note that in fact a table as above represents a set of propositional rules displayed in a vertical manner. Consider a set of propositional rules defined according to the scheme given by (7.2), i.e. each rule is of the form

$$rule\colon \#p_1 \wedge \#p_2 \wedge \ldots \wedge \#p_n \longrightarrow \#h_1 \wedge \#h_2 \wedge \ldots \#h_k,$$

where $\#$ is *nothing* or $\# = \neg$.

As we have mentioned, it is typically assumed that all the rules are somewhat similar, i.e. they use mostly the same preconditions (or their negations) and the same conclusions, however, any two rules are obviously different at least with respect to one symbol. Such a set of rules can be organized and displayed in a nice, regular form of classical *decision table* (Table 7.2).

Table 7.2. Vertical propositional decision table

	rule_1	*rule_2*	...	*rule_m*
p_1	v_{11}	v_{12}	...	v_{1m}
p_2	v_{21}	v_{22}	...	v_{2m}
$\vdots$	$\vdots$	$\vdots$		$\vdots$
p_n	v_{n1}	v_{n2}	...	v_{nm}
h_1	w_{11}	w_{12}	...	w_{1m}
h_2	w_{21}	w_{22}	...	w_{2m}
$\vdots$	$\vdots$	$\vdots$		$\vdots$
h_k	w_{k1}	w_{k2}	...	w_{km}

The main advantage of a table as the presented one follows from a simple and intuitive interpretation. Further, the evaluation procedure can be speeded up due to singular evaluation of conditions during each cycle. Furthermore, tables can be organized in a hierarchical manner, i.e. certain action may require passing the analysis to a lower level i.e. a more specific table (an action similar to the 'go to' instruction in some programming languages).

Decision tables can also be specified *horizontally*, in a way similar to the way of displaying truth-tables in Chap. 1.

In such a case the table provides a single row defining the labels, and below specification of the rules, line by line.

The basic scheme of such table is as follows (Table 7.3).

Table 7.3. The form of horizontal decision table

p_1	p_2	...	p_n	h_1	h_2	...	h_k
v_{11}	v_{12}	...	v_{1n}	w_{11}	w_{12}	...	w_{1k}
v_{21}	v_{22}	...	v_{2n}	w_{21}	w_{22}	...	w_{2k}
$\vdots$	$\vdots$	...	$\vdots$	$\vdots$	$\vdots$	...	$\vdots$
v_{m1}	v_{m2}	...	v_{mn}	w_{m1}	w_{m2}	...	w_{mk}

Now, continuing with the example started in the former section, reconsider the rules specified by (7.27) and (7.28). These rules can be nicely represented in a tabular form. Following is an appropriate decision table (Table 7.4).

The decision table (Table 7.4) provides a complete set of rules in a canonical form for deciding for which interpretations the considered formula

$$(p \Leftrightarrow q) \wedge (r \Leftrightarrow s)$$

is satisfied.

Table 7.4. Decision table for checking formula $(p \Leftrightarrow q) \wedge (r \Leftrightarrow s)$

RuleNo	p	q	r	s	h
0	0	0	0	0	1
1	0	0	0	1	0
2	0	0	1	0	0
3	0	0	1	1	1
4	0	1	0	0	0
5	0	1	0	1	0
6	0	1	1	0	0
7	0	1	1	1	0
8	1	0	0	0	0
9	1	0	0	1	0
10	1	0	1	0	0
11	1	0	1	1	0
12	1	1	0	0	1
13	1	1	0	1	0
14	1	1	1	0	0
15	1	1	1	1	1

7.6.3 Binary Decision Lists

Binary decision lists are perhaps some simplest decision structures incorporating specific rules and some control strategy. They take a simple and intuitive graphical form of a list; such a list specifies a sequence of conditions to be evaluated, coupled with a sequence of actions (or conclusions) to be executed depending on if specific condition in the sequence is true or false.

In order to introduce the generic structure of a decision list, let us consider first a simple, intuitive example. Assume one wants to take lift in order to go to a desired floor. Obviously, he has to carry out a sequence of specific actions; however, the actions and their execution depend on specific conditions.

Without going into very details, the specification of the required sequence of actions may be as follow:

1. Check if you are at the desired floor — if so, stay there; if not — continue.
2. Go to the lift door (unconditional).
3. If the lift is waiting, go inside; if not, press the button and wait.
4. If you are inside, press the desired button and go.
5. If you arrive at certain floor, check if this is the desired floor and if so — go out; if not, continue your travel.

Although the above specification is far from being perfect (e.g. complete) it is enough to show the idea of a decision list. There is one principal line (sequence) of actions, which are conditioned by results of simple checks. Perhaps a more readable form of the above specification can be expressed as a graphical chart (see Fig. 7.1).

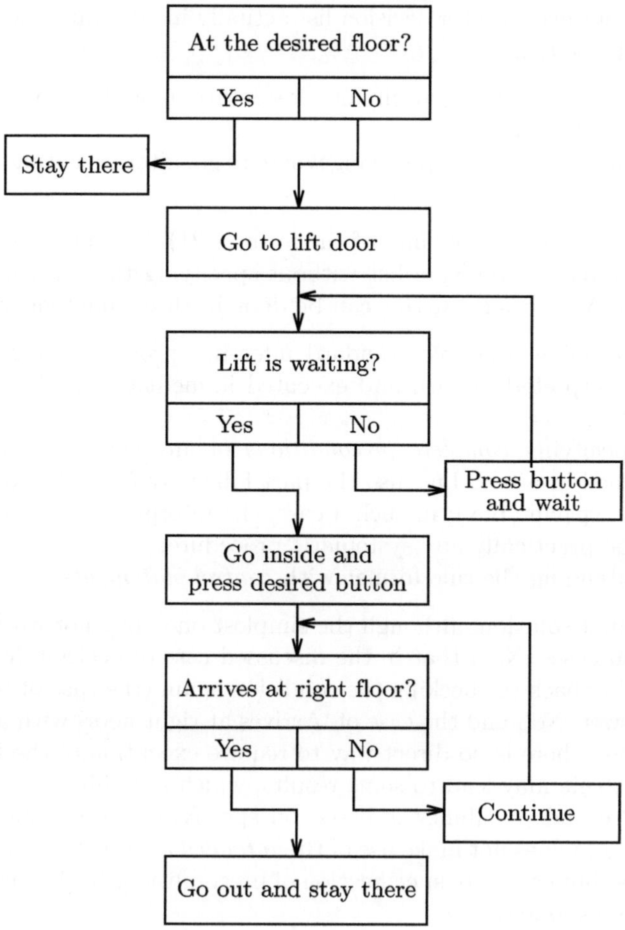

Fig. 7.1. An example decision list

As presented in the picture, the main sequence of actions is the one between checking if one is at the desired floor (the top central box) and going out, and staying at the desired floor (the bottom central box).

It may seem that specifying the logical structure of the rules defined with the decision list is a straightforward task and it is as follows:

$$\begin{aligned} \textit{At the desired floor?} &\longrightarrow \textit{Stay there} \\ \neg(\textit{At the desired floor?}) &\longrightarrow \textit{Go to lift door} \\ \textit{Lift is waiting?} &\longrightarrow \textit{Go inside and press desired button} \\ \neg(\textit{Lift is waiting?}) &\longrightarrow \textit{Press button and wait} \\ \textit{Arrives at right floor?} &\longrightarrow \textit{Go out and stay there} \\ \neg(\textit{Arrives at right floor?}) &\longrightarrow \textit{Continue} \end{aligned} \tag{7.29}$$

Note however, that a decision list actually incorporates two kinds of information; these are:

1) *logical information*, specifying preconditions and actions or decisions in the rule format;
2) *control information*, specifying where to go, after executing or not a specific rule.

Hence, a pure set of rules of the form (7.21), i.e. one as above, cannot, in general, replace a decision list without specifying the control mechanism for the rules. And, basically, this can be done in three different ways:

1) by assuming a simple standard inference procedure, e.g. that the rules are interpreted in turn and executed immediately if their conditions are satisfied;
2) by specifying *complete preconditions* of any rule, i.e. not only the one specified in a box, but also the part following from the context (the path to the specific box); in such a case, the interpretation control mechanism can be practically any systematic procedure;
3) by extending the rule format with *control statements*.

The first solution, although the simplest one, may not work appropriately in certain cases. Note that in the discussed case of decision list there are two loops going back to checking the condition again (the case of 'Lift is waiting?' with answer 'No', and the case of 'Arrives at right floor' with answer 'No'). In such a case there is no direct way to require execution of the loop, and going to a next rule may lead to some results, which are difficult to predict.

The second possibility is based on specifying the complete precondition formula, and does not make use of the *intended context* for a rule. A somewhat simplified but complete specification of preconditions in the discussed example may look as follows:

$$
\begin{array}{rcl}
\textit{At the desired floor?} & \longrightarrow & \textit{Stay there} \\
\neg(\textit{At the desired floor?}) & \longrightarrow & \textit{Go to lift door} \\
\neg(\textit{At the desired floor?}) \wedge \textit{Lift is waiting?} & \longrightarrow & \textit{Go inside} \\
 & & \textit{and press desired button} \\
\neg(\textit{At the desired floor?}) \wedge \neg(\textit{Lift is waiting?}) & \longrightarrow & \textit{Press button and wait} \\
\textit{Arrives at right floor?} & \longrightarrow & \textit{Go out and stay there} \\
\neg(\textit{Arrives at right floor?}) & \longrightarrow & \textit{Continue}
\end{array}
\qquad (7.30)
$$

For simplicity, it is assumed that the case 'Arrives at the right floor?' can be checked only when in a lift, while 'At the desired floor?' only when outside of the lift. This allows for shortening the preconditions.

Specifying full preconditions of the formulae leads to a safe definition of the rules; this is so, since any rule can be fired only when *all* specific conditions enabling activation of the rule are listed explicitly. However, usually such a specification could be very long and in a non-trivial case this approach may turn out to be impractical. In more complex, especially dynamic systems, the validity of preconditions of a rule is limited to being checked only within an intended context, and design of the rule-based system is always hierarchical. Further, in case of dynamic systems, after executing a rule the context changes, and some conditions may turn out not to be verifiable in it.

The third possibility seems to be the most efficient one in case of single-level systems. The details are presented in the following section.

7.6.4 Binary Decision Rules with Control Statements

Consider a simple set of rules given by (7.21) i.e. one specified as below:

$$\begin{aligned} &rule_1\colon \#p_1 \wedge \#p_2 \wedge \ldots \wedge \#p_n \longrightarrow \#h_1, \\ &rule_2\colon \#p_1 \wedge \#p_2 \wedge \ldots \wedge \#p_n \longrightarrow \#h_2, \\ &\vdots \\ &rule_m\colon \#p_1 \wedge \#p_2 \wedge \ldots \wedge \#p_n \longrightarrow \#h_m\,. \end{aligned}$$

In order to provide some control information we shall extend the scheme of the rules with appropriate *Control Statements.* Note that the precondition formula of any rule can play the role of a *filter* or *decision condition.*

Hence, the control information provided with any rule is basically of two types:

1) where to go if the preconditions were satisfied and the rule was fired;
2) where to go if the preconditions formula was not satisfied, and the rule was not fired.

The phrase 'where to go' means in fact pointing to a specific rule. By providing such information a network of rules is specified in fact.

The specification will be done with two keywords: *next*, followed by the number (or identifier) of a rule which should be checked after a specific rule is fired, and *else* followed by the number (or identifier) of a rule which should be checked after a specific rule fails to be fired. A scheme of a single rule now becomes modified as follows

$$rule_i\colon \#p_1 \wedge \#p_2 \wedge \ldots \wedge \#p_n \longrightarrow \#h_1 \;\; next(j) \;\; else(k)\,. \tag{7.31}$$

The above specification of $rule_i$ means that if the rule is fired, the next rule to be checked is $rule_j$, in the other case the next rule to be examined is $rule_k$.

Now we can return to the example concerning using the lift in order to arrive at a desired floor. The complete specification of rules with control statements equivalent to the decision list is as follows:

$$
\begin{array}{llll}
\mathit{rule(1)}: & \mathit{At\ the\ desired\ floor?} \longrightarrow & \mathit{Stay\ there} & \\
 & & \mathit{next()} & \mathit{else(2)}, \\
\mathit{rule(2)}: & \longrightarrow & \mathit{Go\ to\ lift\ door} & \\
 & & \mathit{next(3)} & \mathit{else()}, \\
\mathit{rule(3)}: & \mathit{Lift\ is\ waiting?} \longrightarrow & \mathit{Go\ inside\ and} & \\
 & & \mathit{press\ desired\ button} & \\
 & & \mathit{next(5)} & \mathit{else(4)}, \\
\mathit{rule(4)}: & \longrightarrow & \mathit{Press\ button\ and\ wait} & \\
 & & \mathit{next(3)} & \mathit{else()}, \\
\mathit{rule(5)}: & \mathit{Arrives\ at\ right\ floor?} \longrightarrow & \mathit{Go\ out\ and\ stay\ there} & \\
 & & \mathit{next()} & \mathit{else(6)}, \\
\mathit{rule(6)}: & \longrightarrow & \mathit{Continue} & \\
 & & \mathit{next(5)} & \mathit{else()}.
\end{array}
\tag{7.32}
$$

In the above specification *next*() (empty *next*) means that if the rule is fired, no further rule is specified (perhaps the goal has been reached) and *else*() (empty *else*) means that if the rule fails to be executed no next rule is scheduled; note that the empty *else* part in the above specification occurs in rules with empty preconditions, i.e. ones *always* satisfied. In fact, these are complement rules which specify the action to be done in case the principal action cannot be done directly.

Note also, that through incorporating the explicit control mechanism one no longer has to check the negative (complementary) conditions, which were simply eliminated. This is so, because now the rule operates according to the scheme *if* ⟨*preconditions*⟩ *then* ⟨*conclusions*⟩ *next* ⟨*rule-j*⟩ *else* ⟨*rule-k*⟩, which means that the preconditions are only checked ones, while the *else* part is activated immediately after the check fails.

7.6.5 Binary Decision Trees

Decision trees are useful and intuitive means for specifying various decision procedures. They use simple graphical form of representation (an acyclic, directed graph), and evaluation of decision is a simple matter of traversing such a tree along a given path. There are various forms of decision trees, such as binary decision trees, attributive decision trees, decision trees for analysis of first-order formulae, etc.

Decision trees constitute structures which are *more general* than decision lists. In fact, a decision list is much like a simple decision tree built around a single branch. Decision trees can encode a number of decision lists because they allow *branching*. Decision trees are also *hierarchical structures*, the nodes in a tree are ordered according to a pre-specified hierarchy.

Binary Decision Trees (BDT) are perhaps the best known and most popular decision structures used throughout the widest spectrum of computer science and outside. For intuition, binary decision tree is a structure leading through a series of simple tests to arrive at a certain decision. The tests are limited to evaluation of a propositional statement, which can be *true* or *false*, and hence the trees are referred to as *binary*. In fact, under any decision node (excluding the final decision nodes) there are exactly two branches, one referring to the case when the condition assigned to the node is *true* and the other one referring to the case when the condition is *false*.

Let us introduce a formal definition of *Binary Decision Tree*, which is as follows. Consider a set of propositional symbols P an a set of nodes of the tree, say N. Let $E \subseteq N \times N$ be a binary relation over N.

Definition 76. *A* Binary Decision Tree T *defined on* N, E *and* P *is an acyclic, directed graph satisfying the following properties:*

1. *There is exactly one distinguished node* $n_0 \in N$ *having no parent node (no entry); it is called the* root node *or the* root *of the tree; we shall write* $n_0 = root(T)$.
2. *Any other node has exactly one parent node.*
3. *Any node is assigned a single propositional symbol and it has either two child nodes or none:*
   - *if it has two child nodes, it is a* condition node *and there are two links to child nodes, one referring to the case if the propositional symbol takes* true *and the other one for the case it is* false*;*
   - *if it has no child nodes, it is a* decision node *or a* leaf node*, and the assigned to it propositional symbol indicates the decision.*

Such a binary decision tree is used to specify a decision procedure; it allows to arrive at a decision provided that an appropriate *sequence* of propositional symbols can be evaluated. The tree is traversed top-down, according to some pre-established *order* and at every node the appropriate propositional symbol is evaluated; depending on the result (*true* or *false*) the appropriate branch below is selected, and the next propositional symbol to which the branch leads is evaluated in turn. The final decision is determined after arriving at a leaf node.

Traversing the tree one has to follow a *path* in the tree leading from the root node to one of the leaf nodes. It is useful to define the function of *depth* for any node as follows.

Definition 77. *The depth of a node in a binary decision tree is defined recursively as follows:*

- $depth(n_0) = 0$,
- $depth(n) = depth(parent(n)) + 1$.

In order to show an example of a binary decision tree, let us consider once again example (Subsect. 7.6.1) concerning decision procedure for evaluating propositional formula of the form

$$(p \Leftrightarrow q) \wedge (r \Leftrightarrow s).$$

Recall that we have proposed a set of rules (7.27) and (7.28) as well as an appropriate decision table; now we shall show an appropriate decision tree (Fig. 7.2).

The presented binary decision tree specifies a decision procedure for deciding if the underlying formula given above is satisfied under the interpretation specified with a given path — in fact any path from root to a leaf node defines an interpretation of the formula under discourse. As there are 4 propositional symbols, there are $2^4 = 16$ possible interpretations and there are 16 different paths ended with 16 leaf nodes. For simplicity, 0 defines a branch for which the above proposition is *false* while 1 is the label for *true*.

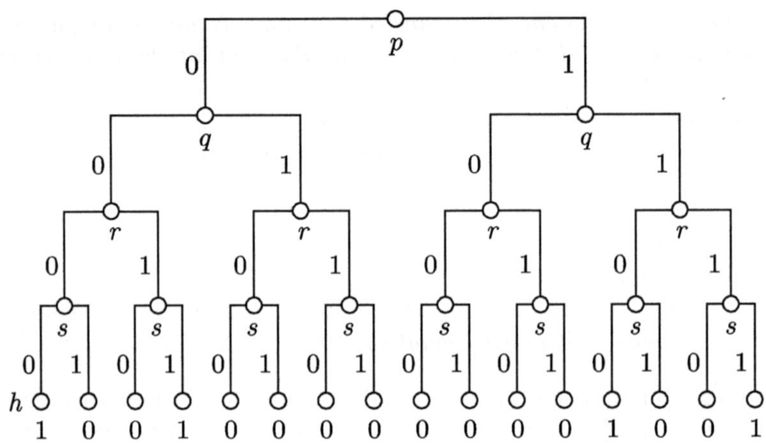

Fig. 7.2. An example binary decision tree for determining the truth-value of formula $(p \Leftrightarrow q) \wedge (r \Leftrightarrow s)$

Recall that decision tree is always a hierarchical decision structure; in fact, any decision tree can be viewed as a structure composed of some smaller trees. Note that any node of a decision tree can be considered as a root note for a *sub-tree* beginning at this node. In this subtree one can distinguish further sub-trees, and so on. In binary tree, below any node (apart from leaf nodes, of course) one may speak about the *left subtree* and the *right subtree*.

Using the concept of a subtree, and knowing that in a binary tree there are only left and right subtrees, the concept of a binary tree can be redefined in a recursive mode as follows.

Definition 78. *A binary decision tree is either:*

- *a single node $n \in N$ assigned unnegated or negated decision defined by propositional symbol $p \in P$,*
- *a graph formed from a node $n_i \in N$ assigned a certain propositional symbol $p \in P$, two subtrees, the left one and the right one, and two arcs labelled with* false *and* true *leading from the node to these subtrees.*

Using the concept of term the two conditions can be restated as follows. Let $t/3$ be a three-argument functional symbol — it will be used as the constructor for the tree. A binary decision tree is a recursive structure defined as follows:

- $t(\_,\_,\#p)$ is a tree,
- if t_{left} and t_{right} are trees, then also $t(t_{left}, t_{right}, p)$ is a tree.

A binary decision tree provides a transparent and intuitive knowledge representation. In fact, the tree shown in Fig. 7.2 is perfectly readable. On the other hand such a graphical representation may be inefficient — some parts of knowledge may be repeated. In fact, in the above example, there are repeated subtrees. For example, the leftmost and rightmost subtree starting at the r nodes (at the depth of 2) are identical. One may ask if the efficiency of representation can be improved. And the answer is yes — the representation with decision tree is not unique, and for a given tree one can find some minimal forms of it.

Note that traversing a decision tree top-down along a certain path is in fact equivalent to evaluating and firing a decision rule. The definition of the rules can be formed when traversing the path in an obvious way. For example, traversing the leftmost path in the tree presented in Fig. 7.2 is equivalent to analyzing a rule of the form

$$\neg p \wedge \neg q \wedge \neg r \wedge \neg s \longrightarrow h.$$

In fact, for any binary decision tree one can always form a set of rules equivalent from logical point of view; the appropriate procedure can be outlined recursively as follows:

1. Start from the root node, and traverse *any* path separately from the root to a certain leaf node; for each path a single separate rule is created.
2. When traversing a path, for any node on that path insert into the precondition formula the propositional symbol assigned to that node; if you select the branch labelled *false*, the propositional symbol should be preceded with negation sign, in case of selecting the branch labelled *true*, the propositional symbol stays unnegated. All the literals forming a path (apart from the one assigned to leaf node) are joined with conjunction.
3. For any path create a rule with the precondition formula defined as above, and conclusion defined by the propositional symbol assigned to the leaf node on the path; if the leaf node is *false*, the conclusion should be preceded with negation sign.

One can check that the set of rules defined with (7.27) and (7.28) is a set of rules logically equivalent to the decision procedure specified with the tree given in Fig. 7.2.

Now consider some two paths leading to two neighboring leaves labelled with 0. The paths leading to them are identical, apart from the last edges — in fact one of them is labelled with 0 (for $\neg s$) and the other one with 1 (for s

without negation). Note however, that disregarding the value taken by s, the decision is the same $h = 0$). Thus it may be no sense in keeping two separate branches leading to the same conclusion — they can be 'glued' together and eliminated. The final decision label is moved up and the tree is simplified.

Consider for example the paths forming the two following rules:

$$\begin{aligned} rule_4 &: \neg p \wedge q \wedge \neg r \wedge \neg s \longrightarrow \neg h, \\ rule_5 &: \neg p \wedge q \wedge \neg r \wedge s \longrightarrow \neg h. \end{aligned} \tag{7.33}$$

Obviously, the two rules have the same effect as a single rule of the form

$$rule_{4-5} \colon \neg p \wedge q \wedge \neg r \longrightarrow \neg h \tag{7.34}$$

in which s do not occur. It can be observed that from logical point of view such reduction of rules is based on application of *backward dual resolution*.

Now assume one repeats the procedure in a recursive way, bottom-up, until no further reduction is possible. Recall that only 'neighboring' nodes can be glued and reduced, and only if they have the same decision value. The reduced in this way decision tree is presented in Fig. 7.3.

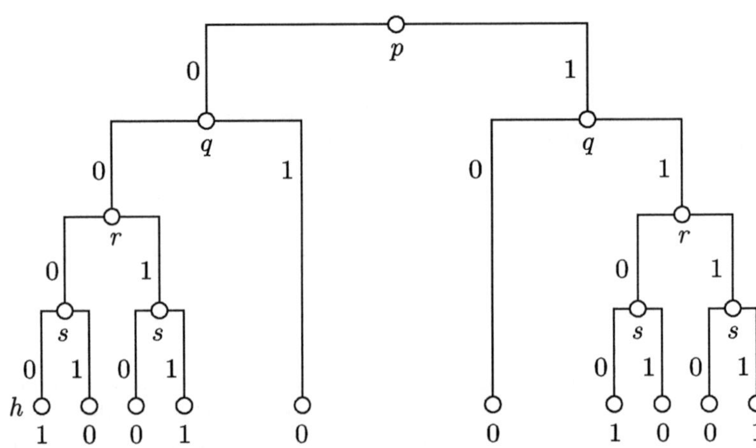

Fig. 7.3. The reduced binary decision tree

Such a reduction procedure has two important features:

1) For a determined order of propositions (and thus the structure of the tree) it always leads to a minimal form when no longer reduction is possible.
2) The obtained in this way decision tree is equivalent to the initial one.

Consider also the reverse procedure of transforming a given set of decision rules into an appropriate decision tree. Let there be given a set of propositional decision rules of the form given by (7.21), i.e.:

$$
\begin{aligned}
&rule_1\colon \#p_1 \wedge \#p_2 \wedge \ldots \wedge \#p_n \longrightarrow \#h_1, \\
&rule_2\colon \#p_1 \wedge \#p_2 \wedge \ldots \wedge \#p_n \longrightarrow \#h_2, \\
&\vdots \\
&rule_m\colon \#p_1 \wedge \#p_2 \wedge \ldots \wedge \#p_n \longrightarrow \#h_m .
\end{aligned}
$$

The following procedure allows to generate a binary decision tree equivalent to the set of rules:

1. Establish an arbitrary *order* of propositional symbols occurring in the preconditions, e.g. $p_1, p_2, \ldots p_n$; proceed from $i = 1$ to $i = n$ according to the order established.
2. For current value of i create a node labelled with p_i and divide the current set of rules into two separate subsets, one where p_i occurs without negation (positive rules) and one where p_i occurs with negation (negative ones). Remove p_i from the precondition of the formulae, and draw two links to child nodes (one for positive rules labelled with *true* and one for negative ones, labelled with *false*). Repeat the procedure in a recursive way until all p_i are assigned to some nodes. The node generated for $i = 1$ is the root node.
3. At the last step, create leaf nodes by assigning literals $\#h_j$ to them.

Since at any step the set of rules is divided into two smaller subsets, and one propositional symbol is removed, the procedure must stop at a certain depth of the tree; in fact, for uniform set of rules as above, the generated tree is of depth n. Every rule is mapped into a single path. Note that the generated tree may have less than 2^n different paths since the number of rules m may be lower than 2^n.

Finally, one may ask a question whether a binary decision tree is *just another* way of specifying a set of propositional decision rules. As we have shown, for any such tree one can easily find an equivalent set of rules, and for any set of rules one can generate an appropriate decision tree. However, there is one important point in the tree generation procedure. It is where one has to establish the *order* of evaluation of the propositional symbols, and in fact it refers to establishing a *hierarchy* in the set of conditions to be evaluated. Hence, one may notice that a tree covers definition of the order among propositional symbols implying the order of evaluation of them in case of interpreting the tree, and influencing the minimal form when reducing the tree. In fact, for a tree built on n propositional symbols, there may be as many as $n!$ different trees (with respect to the order). On the other hand, the order of literals in precondition of a rule is typically ignored, as from logical point of view they form *conjunction* which is commutative.

7.6.6 Binary Decision Diagrams

Binary decision trees presented briefly in the former section constitute an intuitive and very transparent means for describing decision procedures. As it was shown, any such tree can be transformed into equivalent set of rules, so that the rules allow to make the same decision under the same context. One of the disadvantages of decision trees is that knowledge representation with them may be inefficient. As it was shown, repeated subtrees can occupy quite a lot of space. Even reduced trees can be quite large with respect to the encoded information. The Ordered Binary Decision Diagrams (OBDD, for short) [2] constitute perhaps the most concise structures encoding binary decision rules in a graphical form. They can be considered as a certain evolution of classical decision trees.

In order to present the basic idea of Ordered Binary Decision Diagrams let us introduce some initial logical concepts first. After [2] by

$$p \longrightarrow h_0, h_1$$

we shall denote an if-then-else rule of the following form

$$\textit{if } p \textit{ then } h_0 \textit{ else } h_1 \ .$$

From logical point of view such a rule can be considered equivalent to $(p \wedge h_0) \vee (\neg p \wedge h_1)$. Here p is a logical condition, to be called *test expression*, while h_0 and h_1 are conclusions taken in case of p being either *true* or *false*, respectively.

There are two important remarks to be observed. First, this kind of rules can be used as a basic construction for developing any more complex binary decision tree. In fact, p can constitute a label assigned to a branching node, while h_0 and h_1 can be interpreted as the choice of the appropriate left or right subtree.

Second, any logical connective can be redefined with the use of the *if-then-else* operator. For example, denoting *true* with 1 and *false* with 0 we have the following equivalences:

$$p \wedge q \equiv p \longrightarrow q, 0,$$
$$p \vee q \equiv p \longrightarrow 1, q$$

and

$$\neg p \equiv p \longrightarrow 0, 1 \ .$$

Further, for any Boolean formula ϕ the following equivalence allows to reduce the formula by eliminating a single propositional symbol

$$\phi \equiv p \longrightarrow \phi\{p/1\}, \phi\{p/0\} \tag{7.35}$$

where $\phi\{p/t\}$ denotes the formula obtained from ϕ by replacing all the occurrences of p with t. Equivalence (7.35) is known under the name of *Shannon expansion* [2].

By subsequent application of the Shannon expansion any logical formula can be transformed into the so-called *If-then-else Normal Form* (INF), which is a Boolean formula employing the if-then-else operator only. In practice, this means that the formula is transformed into a set of components defining a decision tree for checking satisfiability of the formula.

In order to show that let us apply the Shannon expansion to the formula of Example (Sect. 7.6.1), i.e. the formula given by:

$$\phi = (p \Leftrightarrow q) \wedge (r \Leftrightarrow s).$$

$$\begin{aligned}
\phi &\equiv p \longrightarrow \phi_1, \phi_0 \\
\phi_1 &\equiv q \longrightarrow \phi_{11}, 0 \\
\phi_0 &\equiv q \longrightarrow 0, \phi_{00} \\
\phi_{11} &\equiv r \longrightarrow \phi_{111}, \phi_{110} \\
\phi_{00} &\equiv r \longrightarrow \phi_{001}, \phi_{000} \\
\phi_{111} &\equiv s \longrightarrow 1, 0 \\
\phi_{110} &\equiv s \longrightarrow 0, 1 \\
\phi_{001} &\equiv s \longrightarrow 1, 0 \\
\phi_{000} &\equiv s \longrightarrow 0, 1.
\end{aligned} \tag{7.36}$$

In the above, for shortening the notation, $\phi\{p/0\}$ is denoted as ϕ_0, $\phi\{p/1\}$ is denoted as ϕ_1, etc. Observe that what is generated with the above transformation (7.36) is the specification of the reduced decision tree shown in Fig. 7.3.

The presented above specification has one important feature: it presents *decomposition* of the decision tree into elementary branching 'constructions' (corresponding to subtrees), each of them corresponding to single application of Shannon expansion. Thus it seems relatively easy to identify *identical* subtrees in the main tree — it is enough to compare the appropriate formulae of the ϕ-family. For example, ϕ_{000} is identical to ϕ_{110}, but also ϕ_{00} and ϕ_{11} can be identified to be identical through recursive substitution of identical formulae to the right-hand parts of appropriate equivalences, etc.

The main idea of Binary Decision Diagrams — in contrast to Binary Decision Trees — is that each separate component (subtree) is explicitly displayed only once — repeated occurrences are simply 'glued' together. Since the order of propositional variables is kept identical as in the original tree (in fact, any path in the diagram can be mapped one-to-one to a path in the tree), the diagram is also termed *ordered*.

After identification of identical subtrees the Shannon expansion defined by (7.36) takes the following form:

$$
\begin{aligned}
\phi &\equiv p \longrightarrow \phi_1, \phi_0 \\
\phi_1 &\equiv q \longrightarrow \phi_{11}, 0 \\
\phi_0 &\equiv q \longrightarrow 0, \phi_{11} \\
\phi_{11} &\equiv r \longrightarrow \phi_{111}, \phi_{110} \\
\phi_{111} &\equiv s \longrightarrow 1, 0 \\
\phi_{110} &\equiv s \longrightarrow 0, 1.
\end{aligned}
\tag{7.37}
$$

Figure 7.4 shows an appropriate Ordered Binary Decision Diagram. The negative branches are, by convention, represented with dashed lines, while the positive ones — with solid lines.

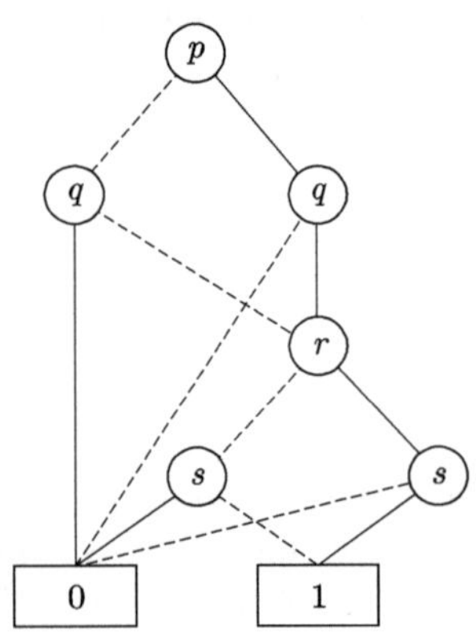

Fig. 7.4. An example Ordered Binary Decision Diagram; case of formula $\phi = (p \Leftrightarrow q) \wedge (r \Leftrightarrow s)$

Note that the representation based on the use of OBDD is much more concise than the one with decision trees; in the presented example, instead of 9 condition nodes in the case of decision tree, there are only six condition nodes in the case of decision diagram. Moreover, 10 final decision nodes were reduced to 2 in case of the diagrams.

Ordered Binary Decision Diagrams can also be *reduced* to a minimal form [2]. It may happen, that an OBDD is non-minimal if one of the two following situations takes place:

1. There are two distinct nodes that have the same propositional symbol assigned and that also have the same low and high-successor subtrees; such nodes can be glued together with their subtrees.
2. There is a node having the same low and high-successor; in this case the test assigned to the node is redundant and it can be removed.

From logical point of view, the first case corresponds to a situation of *physical redundancy*; in fact, for two (or more) nodes, the specification of subtrees rooted at these nodes is *identical*. Practically, such a specification can be encoded with a single model, and the code can be reused if necessary.

The second case means in fact that disregarding the result of the test, one has to proceed the same way; in certain sense the test is redundant and — as such — unnecessary. It can be removed, while its parent node becomes linked directly to common child node.

The two possibilities of reduction are illustrated in Fig. 7.5.

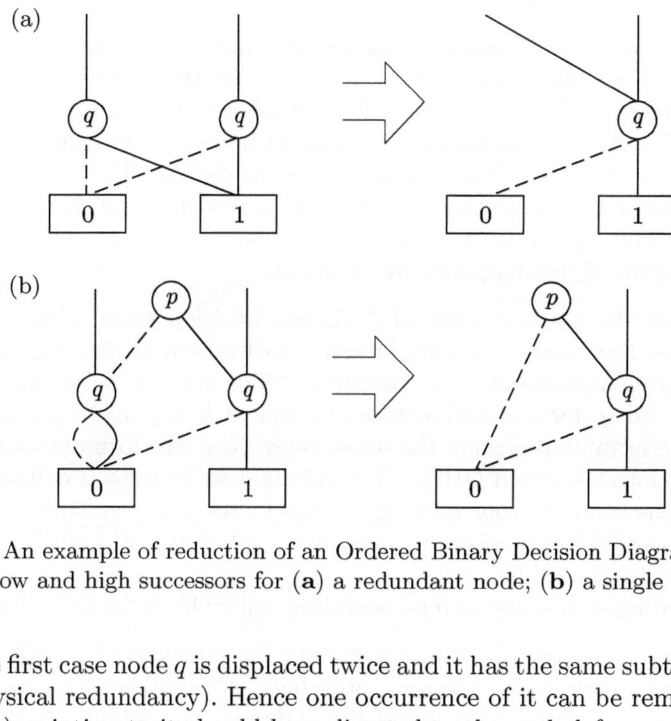

Fig. 7.5. An example of reduction of an Ordered Binary Decision Diagram; case of identical low and high successors for (**a**) a redundant node; (**b**) a single node

In the first case node q is displaced twice and it has the same subtree rooted at it (physical redundancy). Hence one occurrence of it can be removed, and the link(s) pointing to it should be redirected to the node left.

In the second case there are two links from the leftmost node q to the same output. The test at this node is redundant (irrelevant). The node can be removed, and the link pointing to it should be redirected to the output.

Ordered Binary Decision Diagrams constitute perhaps the most concise representation for Boolean formulae and decision procedures based on binary propositional logic. For any Boolean function there exists exactly one reduced OBDD preserving the assumed order of propositional variables. It is stated by the so-called *Canonicity lemma* proved in [2].

Lemma 10. *For any n-argument Boolean function there is exactly one reduced OBDD preserving the assumed order of variables.*

The proof can be found in [2]. The lemma states in fact, that having established the order of propositional symbols, the representation of this function with an OBDD is defined in a unique way. Unfortunately, the size and shape of the OBDD depends on the order of variables (and, needless to say, for n variables there are $n!$ different orderings), and, as shown in [2], in fact the graph of the diagram is sensitive to the ordering to high degree.

Taking into account the above lemma, application of reduced OBDD includes the following activities:

- Checking if a Boolean formula is tautology (is always false); from the above lemma it follows that the appropriate reduced OBDD is just terminal node 1 (0).
- Checking if a Boolean formula is satisfiable; in fact, it is enough to see if at least one path in the appropriate OBDD leads to terminal node 1. Moreover, all the interpretations satisfying the formula are defined with appropriate paths leading to terminal node 1 — in order to find them it is enough to search for all such paths in the OBDD.
- Comparing two (or more) formulae: in order to say if two formulae are logically equivalent it is enough to transform them to reduced OBDD and compare if the diagrams are identical.

From the point of view of designing decision structures, and especially rule-based systems, the OBDD can also be used to determine some minimal representations of such structures. Since any set of decision rules can be considered to form a conjunction of them, it is relatively straightforward to find the formula replacing the set of rules. Next the formula should be transformed into a reduced OBDD. The minimal set of rules is defined then by all the paths from the root node to terminal nodes. Alternatively, the initial set of rules could be translated into a binary decision tree, and the tree could be reduced to an OBDD.

Finding such minimal representation suffers from the following limitations:

- The final minimal form depends on the established ordering of propositional symbols; for different orderings, different minimal forms (also different with respect to size of the diagram) can potentially be found.
- The transformation to an OBDD (reduced OBDD) itself is a complex computational procedure.
- This approach is limited to the case of propositional logic and a single yes-no output.

Hence, although OBDD constitute an interesting concept from theoretical point of view, their practical applications in the domain of knowledge engineering seems to be intrinsically limited. However, certain ideas concerning both *ordering* and *reduction* can be influencing for some more reach systems with special attention paid to ones incorporating attribute-based logics.

7.7 Dynamic and Non-Monotonic Systems

In case of more complex applications it may be the case that *dynamic changes* of knowledge base are necessary to model the current state of knowledge. This means that the inference process is no longer a *monotonic* one. New facts are concluded to be true, while some other are no longer true, and as such, they have to be deleted from the database. This may happen in the case of modeling dynamic systems behavior or while performing non-monotonic reasoning in case of incomplete initial information.

In case of dynamic rule-based systems the scheme of the rules is no longer as simple as that of (7.1) or (7.2). In order to model dynamic changes in the knowledge base two basic operations are necessary; these are:

1) *retract(q)* — which retracts (deletes) fact q from the knowledge base,
2) *assert(q)* — which asserts (adds) fact q to the knowledge base.

The basic scheme of a propositional rule which enables dynamic changes of the knowledge base is as follows

$$rule\colon p_1 \wedge p_2 \wedge \ldots \wedge p_n \longrightarrow retract(d_1, d_2, \ldots, d_d), assert(h_1, h_2, \ldots, h_h). \quad (7.38)$$

As in the former case of classical monotonic propositional rules given by (7.1) and (7.2) the $LHS(rule) = p_1 \wedge p_2 \wedge \ldots \wedge p_n$ is a conjunction of propositional literals; each p_i may be a positive atom or a negated one. The $LHS(rule)$ part of the rule defines its *preconditions*, i.e. conditions which must simultaneously hold to fire the rule. On the other hand, now the $RHS(rule) = retract(d_1, d_2, \ldots, d_d), assert(h_1, h_2, \ldots, h_h)$ and it defines the changes to be executed on the knowledge base if the rule is fired.

One typical application of such rules concerns modeling dynamic systems with memory. As an example consider the well known *SR* trigger. It has two binary inputs s (set) and r (reset) and one output q being equal to the state of internal memory. The logical function describing its behavior is given by the following equation

$$q' = q \wedge \neg r \vee s.$$

The dynamic behavior of the trigger can be described with the following set of rules:

$$rule_1\colon q \wedge r \wedge \neg s \longrightarrow retract(q), assert(\neg q),$$
$$rule_2\colon \neg q \wedge \neg r \wedge s \longrightarrow retract(\neg q), assert(q)\ .$$

The interpretation of the rules is as follows. Rule$_1$ says: *if q is set to 1 (true), then it is reseted by r under absence of s.* Rule$_2$ says: *if q is set to 0 (false), then it is set by s under the absence of r.* As there are no more rules, under any other combination the memory stays unchanged.

8

Rule-Based Systems in Attributive Logic

Although the basic concepts of rule based systems can be nicely explained at the level of propositional logic — as it was done in Chap. 7 — in case of more realistic, practical applications propositional logic becomes too poor as a language for knowledge encoding. Obviously, expressive power of propositional languages is far too low for efficient knowledge representation. Hence, since a more powerful language is necessary, one of the most obvious choices is employing a wide spectrum of attributive languages.

The languages basing on the concept of attributes have a number of characteristics, making them almost ideal tools for practical representation and manipulation of knowledge; these include the following features:

- **introducing variables** — attributes play the role of *variables*; the same attribute can take different values and there is no need to introduce new propositional symbols;
- **easy specification of constraints** — since attributes play the role of variables, using various relational symbols allows to specify almost any constraints;
- **parameterization** — attributes may also play the role of parameters to be instantiated at some desired point of inference.

The above characteristics contribute to **increased expressive power** — as the result of introducing attributes (as variables) and domains for them. Nevertheless, attributive languages stay intuitive and transparent.

In Chap. 3 we have introduced four different types of logical languages based on the use of attributes; these are:

- AAL — *Atomic Attributive Logic*, i.e. attributive logic with atomic values of attributes only.
- SAL — *Set Attributive Logic*, i.e attributive logic with set values of attributes.
- VAAL — *Variable Atomic Attributive Logic*, i.e. attributive logic with atomic values of attributes incorporating variables.

A. Ligęza: *Logical Foundations for Rule-Based Systems*, Studies in Computational Intelligence (SCI) **11**, 129–153 (2006)
www.springerlink.com

- VSAL — *Variable Set Attributive Logic*, i.e attributive logic with set values of attributes incorporating variables.

Obviously, there are many other possibilities of enhancing the expressive power and knowledge representation capabilities. Some most frequently used options include, but are not limited to, the following features:

- admission of various relational symbols, such as $<$, $\leq$, $\in$, $\subseteq$, etc.;
- explicit specification of type of attributes (variables), e.g. logical (yes/no) values, nominal sets, integers, real numbers, etc.;
- incorporation of general and domain specific operations on attributes and their values (e.g. calculation of functions).

In this Chapter we present various knowledge representation forms of rule-based systems based on attributive logic. In particular, special attention is paid to the following four most frequently used or most promising forms:

- Attributive Decision Tables (AD-Tables) and Extended Attributive Decision Tables (XAD-Tables),
- Attributive Decision Trees (AD-Trees) and Extended Attributive Decision Trees (XAD-Trees),
- Tabular Trees (Tab-Trees, TT), and Extended Tabular Trees (XTT),
- Attributive Rule-Based Systems.

Attributive Decision Tables and Attributive Decision Trees constitute some extensions of their propositional prototypes. Various forms of them are used in information systems [106], machine learning [18], and other domains.

A concept of two-level, hierarchical tabular rule-based system was introduced in [65]; the upper-level rules were devoted to switching to lower-level tables. The rules of a lower-level table were destined to operate within a given context.

Tabular-Trees constitute an extension of this idea and simultaneously an interesting combination of decision trees and attributive decision tables; they were first explicitly proposed in [141], and further discussed in [97]. They seem to constitute an intuitive and transparent conceptual tool for highly efficient presentation of rule-based knowledge in a hierarchical way which seems to be most promising.

Recent developments [92] resulted in an elaborated version of Extended Tabular Trees (XTT)[1] and new graphical software tool called MIRELLA for visual edition and analysis of attributive tabular rule-based systems.

8.1 Attributive Decision Tables

One of the main disadvantages of decision tables [120] or ones defined as by Table (7.2) is the consequence of very limited possibilities of defining the pre-

[1] The name *eXtended Tabular Trees* first appeared in [92] and is due to Grzegorz Jacek Nalepa.

conditions of actions with the use of binary values only, while in numerous cases the use of values of attributes is much more convenient. A table incorporating attributes and providing specification of their values is an *Attributive Decision Table* (AD-Table). Such tables are also called *Object-Attribute-Value Table* (OAV Table, OAT) or decision tables, as well. In extended form, allowing for non-atomic values of attributes they are called *Extended Attributive Decision Tables* (XAD-Tables).

Below the AD-Tables and their extended form being XAD-Tables are presented in turn.

8.1.1 Basic Attributive Decision Tables

The AD-Tables in their basic form are very similar to relational database tables, and in fact it is a RDB table with specific interpretation of certain columns. The basic scheme of such a table is presented below (Table 8.1). Here *att* stays for *attribute* and *con* for *conclusion*.

Table 8.1. Basic scheme of an attributive decision table

att_1	*att_2*	...	*att_k*	*con_1*	*con_2*	...	*con_m*
v_{11}	v_{12}	...	v_{1k}	w_{11}	w_{12}	...	w_{1m}
v_{21}	v_{22}	...	v_{2k}	w_{21}	w_{22}	...	w_{2n}
⋮	⋮		⋮	⋮	⋮		⋮
v_{n1}	v_{n2}	...	v_{nk}	w_{n1}	w_{n2}	...	w_{nm}

In the above table, the rows specify under what attribute values certain conclusion may be drawn (or an action may be executed). Since both v_{ij} and w_{ij} may take several values (not just true/false, or unimportant), the approach is more general than the former one, based on classical tables. In case of actions, the specific values may indicate if the action is to be applied or not, but they can also specify a certain *parameter* of the action associated with specific column.

The interpretation (and execution) of this table is straightforward: the rows are examined in turn, and the current values of subsequent attributes are determined; if they match the pattern specified in the examined row, the actions specified in the right-hand part of the table are executed, and next row is examined in turn. Of course, *hierarchical organization* of tables is also possible.

The rules covered by a decision table are put all together and form a kind of *decision unit* designed to work in a certain assumed context situation. The conditional part of a table can also be empty — in such a case the table specifies knowledge which is unconditionally true; this can be interpreted as facts (from logical point of view) or as relational database tables describing certain objects.

8.1.2 Information Systems

In his seminal book, Pawlak [106] introduces the concept of *information systems* or *attribute-value system*, also determined as *Knowledge Representation System* (or *KR-system* for short). Let us recall some basic definitions.

Let U denote a finite, nonempty set (of objects), to be called the *universe*. Further, let A denote a finite set of primitive[2] attributes. Every (primitive) attribute $A_i \in A$ is a function of the form

$$A_i \colon U \rightarrow D_i \ ,$$

where D_i is the set of legal values of A_i, to be called the *domain* of A_i.

Definition 79 (Knowledge Representation System [106]). *A Knowledge Representation System (KR-System) is a pair* $S = (U, A)$.

Usually, the specification of a KR-System can be presented in the tabular form, similar to the one of relational databases. This implies that the value of *every* attribute for *every* object of U is given in an explicit way.

In his book, Pawlak [106] gives also a formal definition of a decision table. Let $K = (U, A)$ be a knowledge representation system, and let $C \subseteq A$ and $H \subseteq A$ be two subsets of the set of all attributes, to be called *conditional* (*condition*) and *decision* attributes. A KR-system with distinguished conditional and decision attributes is called a *decision table*.

Definition 80 (Decision Table (Pawlak) [106]). *A decision table is a four-tuple* $T = (U, A, C, H)$.

Note that, according to the above definition, a decision table has conditional part (the one with conditional attributes C) and decision (conclusion) part (the one described with decision attributes H). Although not stated explicitly, both the definition of KR-system and the one of decision table seem to allow only atomic values of attributes.

For the purpose of this book we shall introduce the following notation and definition of XD-Tables, which, basically, are very much the same, as given by Definition 80.

Consider a nonempty, finite set of attributes of interest, $A = \{A_1, A_2, \ldots, A_n\}$. For any attribute A_i let D_i denote the domain of this attribute, $i = 1, 2, \ldots, n$. The domain can be a finite one, i.e. $D_i = \{d^1, d^2, \ldots, d^{m_i}\}$, or infinite, e.g. $D_i \subseteq \mathbb{R}$, where $\mathbb{R}$ is the set of real numbers.

Attributes $A_1, A_2, \ldots, A_n$ denote some properties of interest, selected for expressing the domain knowledge of the analyzed system, when operating in a specific (local) context. They are aimed at representation of precondition

[2] Pawlak [106] distinguishes between *primitive* attributes, which are here called simply *attributes*, and sets of them (subsets of A) which he calls *compound* attributes, to be called here also compound or *meta* attributes.

knowledge for the rules. It is typically implicitly assumed that the attributes are independent from one another.

Further, let us consider a specific set of attributes $H = \{H_1, H_2, \ldots, H_m\}$ with the domains $D_1^H, D_2^H, \ldots, D_m^H$ where $D_h^H = \{h^1, h^2, \ldots, h^{m_h}\}$. These attributes are aimed at describing the output of the rules, e.g. conclusions, decisions or other output values.

In the basic statement, the structure of a single condition (atomic formula) is as simple as an atomic formula of the AAL language, see Definition 49. Hence, if $u \in U$ denotes an object, $A_i \in A$ is an attribute and $d \in D_i$ is an atomic value of the domain of A_i, any expression of the form

$$A_i(u) = d$$

is an atomic formula. The decision statements are constructed in a similar way, e.g. $H_j(u) = h$.

Tabular system can be used both for data templates representation and for representation of rules. In the former case no decision attributes H are present[3]; the particular rows of the table represent formulae describing individual items. In case of rules, specific columns with the decision (conclusion) attributes H are present, and the other attributes play the role of variables used in the specification of preconditions.

In general, the tabular representation follows the pattern of RDB systems. They can be used to represent both *Data* and *Knowledge* (*D&K*), so they provide a common representation for those two classes of information [61, 62]. Below a brief note on the differentiation of *data* and *knowledge* is presented.

What is Data

An **atomic data item** is a certain piece of information represented in certain accepted language, and is:

- as precise as possible (within the selected language),
- meaningful (having some interpretation),
- positive (no negation is used),
- unconditional.

Examples of data items include: propositional formula, ground atomic formula of predicate calculus, O-A-V fact, PROLOG fact, etc.

A **data item** is a conjunctive combination of atomic data items. Examples of data items include: ground conjunctive formulae, records (of atomic data items) in RDB, etc.

Data is a collection of data items. Examples include RDB system tables, collections of ground conjunctive formulae, etc.

---

[3] To be consistent, we should perhaps use the attributes of H to denote unconditional statements; here we selected A, used traditionally also in databases.

Data and knowledge can also be differentiated by their intended interpretation: a data item (such as an attribute value, record, table) is considered to be *data* if the main intended use of it is to provide static, detailed and precise image of a fragment of real world while a knowledge item (such as fact, simple conjunctive formula, DNF formula, and especially rules) is intended to provide more general knowledge defining universal or local properties of the world. From practical point of view, one can consider data to be the part of knowledge that is unconditional and expressed with the *finest granularity*.

What is Knowledge

An **atomic knowledge item** is any data item and any more general elementary item of the accepted language, which:

- may contain variables/sets/intervals/structures (according to the selected language),
- meaningful (having some interpretation),
- positive or negative,
- perhaps conditional.

Examples of atomic knowledge items include: atomic formula of predicate calculus, extended O-A-V fact[4], PROLOG fact with variables, etc.

A **knowledge item** is a conjunctive combination of atomic knowledge items. Examples of knowledge items include: conjunctive formulae, records (of atomic knowledge items), etc.

Knowledge is a collection of knowledge items. Examples include relational database-like tables specifying data templates or rules, decision tables, collections of logical formulae, PROLOG programs.

If the specification contains variables (e.g. universally quantified, or defining some scope ones) or it is true only under certain conditions (e.g. takes the form of rules, allows for deduction or any other form of inference), then it should be normally considered to be *knowledge*. However, in the uniform, simplified model proposed in this book explicit distinction is in fact not necessary. A RDB table would be normally considered as data, but it may be considered as most detailed knowledge as well. On the other hand, tabular system of data templates can be considered as extensional specification of data.

8.1.3 Attributive Decision Tables with Atomic Values of Attributes

Consider a set of rules, each of the form:

$$r_i\colon (A_1 = d_{i1}) \wedge (A_2 = d_{i2}) \wedge \ldots (A_n = d_{in}) \longrightarrow$$
$$\longrightarrow H_1 = h_{i1} \wedge H_2 = h_{i2} \wedge \ldots H_m = h_{im}.$$

[4] O-A-V stays for Object-Attribute-Value.

Now, taking into account the advantage of the uniform form of all the rules in the system, the set of rules can be specified in a transparent, tabular form (one resembling database format). The definition of (a basic form of) an AD-Table follows.

Definition 81 (Attributive Decision Table). *An AD-Table is a table of the following form:*

$$\boldsymbol{T} = \begin{array}{c|ccccc|cccc}
\hline
rule & A_1 & A_2 & \ldots & A_j & \ldots & A_n & H_1 & H_2 & \ldots & H_m \\
\hline
r_1 & d_{11} & d_{12} & \ldots & d_{1j} & \ldots & d_{1n} & h_{11} & h_{12} & \ldots & h_{1m} \\
r_2 & d_{21} & d_{22} & \ldots & d_{2j} & \ldots & d_{2n} & h_{21} & h_{22} & \ldots & h_{2m} \\
\vdots & \vdots & \vdots & & \vdots & & \vdots & \vdots & \vdots & \vdots & \vdots \\
r_i & d_{i1} & d_{i2} & \ldots & d_{ij} & \ldots & d_{in} & h_{i1} & h_{i2} & \ldots & h_{im} \\
\vdots & \vdots & \vdots & & \vdots & & \vdots & \vdots & \vdots & \vdots & \vdots \\
r_k & d_{k1} & d_{k2} & \ldots & d_{kj} & \ldots & d_{kn} & h_{k1} & h_{k2} & \ldots & h_{km} \\
\hline
\end{array} \qquad (8.1)$$

The above table represents k uniformly structured decision rules.

Using the matrix notation one can also write $\mathbf{T} = [\mathbf{R\Phi H}]$, where $\mathbf{R}$ is the leftmost column vector of rule names, $\mathbf{\Phi}$ is the matrix of conditional attribute values, and $\mathbf{H}$ is the rightmost matrix specifying conclusions (decisions). Further, the precondition matrix $\mathbf{\Phi}$ can be (logically) written as $\Phi = \phi_1 \vee \phi_2 \vee \ldots \vee \phi_m$, or in the matrix form as

$$\mathbf{\Phi} = \begin{bmatrix} \phi_1 \\ \phi_2 \\ \vdots \\ \phi_k \end{bmatrix},$$

where $\phi_i = (A_1 = d_{i1}) \wedge (A_2 = d_{i2}) \wedge \ldots \wedge (A_n = d_{in})$. From now on Φ will denote the logical formula corresponding to tabular form $\mathbf{\Phi}$ and vice versa.

8.1.4 Example: Opticians Decision Table

After [106][5] let us consider the following set of attributes for Opticians Decision System:

- $A_1 :=$ age; $D_1 = \{y, p, q\}$, where:
  y — young,
  p — pre-presbyotic,
  q — presbyotic;
- $A_2 :=$ spectacle; $D_2 = \{m, h\}$, where:
  m — myope,
  h — hypermyope;

[5] This example is originally due to Cendrowska, see [15].

- A_3 := astigmatic; $D_3 = \{n, y\}$, where:
  n — no,
  y — yes;
- A_4 := tear production rate; $D_4 = \{r, n\}$, where:
  r — reduced,
  n — normal;
- D := type of contact lenses (decision attribute); $D_D = \{H, S, N\}$, where:
  H — Hard contact lenses,
  S — Soft contact lenses,
  N — No contact lenses.

Consider the Opticians Decision Table (Table 8.2), being a perfect example of an AD-table.

Table 8.2. Optician Decision Table

Number	Age	Spectacle	Astigmatic	Tear p.r.	Decision
1	y	m	y	n	H
2	y	n	y	n	H
3	p	m	y	n	H
4	q	m	y	n	H
5	y	m	n	n	S
6	y	n	n	n	S
7	p	m	n	n	S
8	p	n	n	n	S
9	q	n	n	n	S
10	y	m	n	r	N
11	y	m	y	r	N
12	y	n	n	r	N
13	y	n	y	r	N
14	p	m	n	r	N
15	p	m	y	r	N
16	p	n	n	r	N
17	p	n	y	r	N
18	p	n	y	n	N
19	q	m	n	r	N
20	q	m	n	n	N
21	q	m	y	r	N
22	q	n	n	r	N
23	q	n	y	r	N
24	q	n	y	n	N

For intuition, this table can be regarded as a *complete* decision table since for every possible combination of input attribute values there is a decision provided. Further, the provided decision is unique in any case — the system is said to be deterministic. The table can be viewed as a specification of a tabular rule based system. A brief analysis assures us that there are no redundant or subsumed rules. We shall return to precise analysis of the system later on, after defining formal properties of rule-based systems.

8.2 Extended Attributive Decision Systems

Encoding decision tables with the use of atomic values of attributes only simplifies design and analysis of such systems. Any two rules can be easily compared (e.g. if they are identical), and if the attributes are really functional (for any object or input situation they take only a single value). Two rules with different preconditions are sure to be deterministic (either first or the second can be fired, but not both of them together). Moreover, for some analyzes and operations on such systems one can apply simple relational algebra, as in the case of relational databases [23].

Unfortunately, for a number of more complex, realistic applications, it is not enough to stay at the level of atomic values only. There are at least the following three reasons for that:

1) many attributes are in fact generalized attributes (pseudo-functions) — they can take several values (a subset) from the domain at a certain instant of time, i.e. they are mappings of the form $A_i : U \to 2^{D_i}$ (see Def. 48);
2) knowledge specification (both in case of rules and facts) with non-atomic values can be far more concise than in the case of atomic values;
3) it is often impossible to specify precise definition of preconditions, and one needs to have the possibility to specify attribute values as belonging to intervals, subset of the domain, etc.

The main extension with respect to AD-Tables with atomic values consists then of admitting non-atomic values of attributes, both in the conditional part and in the conclusion part.

In principle, the value of an attribute can be any *subset* of its domain[6], or in general an element of a lattice. In case of sets we shall assume that only finite domains will be considered. Hence, instead of a set of real numbers one has to choose a discrete (and bounded) subset of it; since the approach is computer-oriented, this assumption does not seem to be too restrictive.

The set domains can be of the following types:

- **nominal sets** — unordered sets of discrete elements,
- **ordered sets** — ordered sets of discrete elements,
- **lattice** — the sets with partial order relation forming a lattice structure.

In the class of ordered sets it is typical to distinguish a set of numbers, e.g. a finite subset of integer numbers or real ones. Typically also one distinguishes attributes taking logical values, such as *true* and *false* (also denoted with 1 and 0).

---

[6] For simplicity, we do not allow here any structures, such as terms in FOPC.

Now consider a basic fact of the SAL language defined by Definition 50, i.e. one defined as below. Let $u \in U$ be an object, $A_i \in A$ be an attribute and let $t \subseteq D_i$ be some subset of the domain of A_i. Two most typical atomic formulae will be considered, i.e. the one with equation of the form

$$A_i(u) = t$$

and the one with 'belongs to' relation

$$A_i(u) \in t.$$

Both of the formulae are atomic formulae of SAL. On the base of atoms, appropriate selectors can be constructed to form preconditions of the rules.

Consider a set of rules, each of the form

$$r_i\colon\ (A_1 \in t_{i1}) \wedge (A_2 \in t_{i2}) \wedge \ldots (A_n \in t_{in}) \longrightarrow \\ \longrightarrow H_1 = h_{i1} \wedge H_2 = h_{i2} \wedge \ldots H_m = h_{im}.$$

Taking into account the advantage of the uniform form of all the rules in the system, the set of rules can be specified in a transparent, tabular form (one resembling database format). The definition of (a basic form of) a XAD-Table follows.

Definition 82 (Extended Attributive Decision Table). *An XAD-Table is a table of the following form:*

$$T = \begin{array}{c|ccccc|cccc} \hline \textit{rule} & A_1 & A_2 & \ldots & A_j\ \ldots & A_n & H_1 & H_2 & \ldots & H_m \\ \hline r_1 & t_{11} & t_{12} & \ldots & t_{1j}\ \ldots & t_{1n} & h_{11} & h_{12} & \ldots & h_{1m} \\ r_2 & t_{21} & t_{22} & \ldots & t_{2j}\ \ldots & t_{2n} & h_{21} & h_{22} & \ldots & h_{2m} \\ \vdots & \vdots & \vdots & & \vdots & \vdots & \vdots & \vdots & \vdots & \vdots \\ r_i & t_{i1} & t_{i2} & \ldots & t_{ij}\ \ldots & t_{in} & h_{i1} & h_{i2} & \ldots & h_{im} \\ \vdots & \vdots & \vdots & & \vdots & \vdots & \vdots & \vdots & \vdots & \vdots \\ r_k & t_{k1} & t_{k2} & \ldots & t_{kj}\ \ldots & t_{kn} & h_{k1} & h_{k2} & \ldots & h_{km} \\ \hline \end{array} \qquad (8.2)$$

The above table represents k uniformly structured decision rules with non-atomic values of attributes. The matrix notation can be also used as in the case of AD-Tables.

The interpretation of set values of attributes is as follows. Let t be a subset of the domain of attribute A_i, $t = \{d_1, d_2, \ldots, d_j\}$. In preconditions, a selector like

$$A_i \in t$$

can be expressed as

$$A_i = d_1 \vee A_i = d_2 \vee \ldots \vee A_i = d_j,$$

i.e it denotes the so-called *internal disjunction* (see Subsect. 3.5.2 on page 60). For the condition to be satisfied it is enough that $A_i = d$, where $d \in t$. In case of attributes taking set values, it can also happen that $A_i = t'$ and for satisfaction of the condition it is enough that $t' \cap t \neq \emptyset$.

On the other hand, in decision part of the XAD-table, an expression of the form

$$H_i = t$$

means in fact *internal conjunction* (see Subsect. 3.5.1 on page 59), i.e. it can be interpreted as

$$H_i = d_1 \wedge H_i = d_2 \wedge \ldots \wedge H_i = d_n .$$

This interpretation follows from practical considerations — the preconditions of the rule are usually aimed at specifying some wider context of application, while its conclusions are as precise as necessary.

8.3 Example

Let us continue with the Opticians Decision Table presented in Subsect. 8.1.4. It is not difficult to see, that the table can be represented in equivalent but shorter form as follows (Table 8.3).

Table 8.3. A reduced form of the Optician Decision Table

Number	Age	Spectacle	Astigmatic	Tear p.r.	Decision
1	y	—	y	n	H
2	—	m	y	n	H
3	y	—	n	n	S
4	p	—	n	n	S
5	—	n	n	n	S
6	—	—	—	r	N
7	p	n	y	—	N
8	q	m	n	—	N
9	q	n	y	—	N

Moreover, even a more concise specification is possible (Table 8.4). The final table is still equivalent to the initial one.

8.4 Attributive Rule-Based Systems

In this Section a general format of attributive decision rules for construction of rule-based systems will be presented. As usual, consider a nonempty, finite set of attributes of interest, $A = \{A_1, A_2, \ldots, A_n\}$. For any attribute A_i let D_i

Table 8.4. The most reduced form of the Optician Decision Table

Number	Age	Spectacle	Astigmatic	Tear p.r.	Decision
1	y	—	y	n	H
2	—	m	y	n	H
3–4	$\{y, p\}$	—	n	n	S
5	—	n	n	n	S
6	—	—	—	r	N
7–9	$\{p, q\}$	n	y	—	N
8	q	m	n	—	N

denote the domain of this attribute, $i = 1, 2, \ldots, n$. The domain, as indicated before, can be a finite one, i.e. $D_i = \{d^1, d^2, \ldots, d^{m_i}\}$, or infinite, e.g. $D_i \subseteq \mathbb{R}$, where $\mathbb{R}$ is the set of real numbers.

8.4.1 Rule Format

The most general form of a rule considered here is an extended form of the rules discussed in Sect. 7.7. It is based on the basic rule format but includes both *control statement* and *dynamic operations* definitions. Hence the rule can operate on system memory and show where to pass control in an explicit way.

Recall that in the case of propositional rules, the rule incorporating control statement were specified by equation (7.31) as follows

$$rule_i \colon \#p_1 \wedge \#p_2 \wedge \ldots \wedge \#p_n \longrightarrow \#h_1 \;\; next(j) \;\; else(k),$$

i.e. with the use of the *next(j)* part specifying which rule should be examined immediately after successful execution of rule i, and with the *else(k)* part specifying which rule should be tried in case of failure.

Moreover, the specification of dynamic operations was given by the *retract* and *assert* parts by expression (7.38), and it took the following form

$$rule \colon p_1 \wedge p_2 \wedge \ldots \wedge p_n \longrightarrow retract(d_1, d_2, \ldots, d_d), assert(h_1, h_2, \ldots, h_h).$$

Now, the extended form of a rule in attributive languages may, in general, incorporate the following components:

- *a unique identifier* of the rule; it can be the name or the number of the rule, or both;
- *context specification* of a rule; it is a form of general preconditions, specifying a context in which the rule is expected to be meaningful; in fact, contexts can be specified in a hierarchical way, but for simplicity only one level of contexts is considered here;
- *preconditions* of the rule, specifying logical formula that has to be satisfied in order for the rule to be executed;

- *dynamic operation specification* with the use of the *retract* and *assert* parts;
- *conclusion, decision* or *action* part being the final output of the rule;
- *control specification* with the use of the *next* and *else* parts.

The generic form of such a rule can be presented as follows:

$$\begin{aligned}
rule(i)\colon\ & \psi \wedge \\
& A_1 \in t_1 \wedge A_2 \in t_2 \wedge \ldots \wedge A_n \in t_n \\
& \longrightarrow \\
& retract(B_1 = b_1, B_2 = b_2, \ldots, B_b = b_b) \\
& assert(C_1 = c_1, C_2 = c_2, \ldots, C_c = c_c) \\
& H_1 = h_1, H_2 = h_2, \ldots, H_h = h_h \\
& next(j),\ \ else(k).
\end{aligned} \tag{8.3}$$

In the above specification:

- ψ is a formula defining the context in which the rule is designed to operate; the context may also be specified outside for a group of rules forming a decision table,
- $A_1 \in t_1 \wedge \ldots \wedge A_n \in t_n$ is the regular precondition formula,
- $B_1 = b_1, \ldots, B_b = b_b$ is the specification of the facts to be retracted from the knowledge base,
- $C_1 = c_1, \ldots, C_c = c_c$ is the specification of the facts to be asserted to the knowledge base,
- $H_1 = h_1, \ldots, H_h = h_h$ is the specification of conclusions forming a direct output of the rule (e.g. decisions to be displayed on the terminal or control actions to be executed),
- $next(j)$, $else(k)$ are the specifications of control.

In real rules some of the attributes in the precondition of the rule may be omitted; this is equivalent to specifying conditions of the form $A_i \in \_$. The specification of values of attributes $B_1 = b_1, B_2 = b_2, \ldots, B_b = b_b$, $C_1 = c_1, C_2 = c_2, \ldots, C_c = c_c$ and $H_1 = h_1, H_2 = h_2, \ldots, H_h = h_h$ concern attributes defined for the system, i.e. $B_1, B_2, \ldots, B_b, C_1, C_2, \ldots, C_c$, $H_1, \ldots, H_h \in A$.

A general format of extended rule specification is also discussed in details in Sect. 9.4.

8.4.2 Rule Firing

Examination and possible firing of a rule given by (8.3) is performed by the rule interpreter according to the following specification:

1. As an initial condition, it must be checked if the context defining formula ψ is satisfied — if so, proceed; if not then there is no use in further examining the rule. The system must determine the current context and switch to the group of rules designated to operate in this context. Note that there are two basic possibilities of context checking:

- checking the context any time a rule is considered for execution; we shall refer to such an approach as *permanent context checking*;
- checking the context once when entering a group of rules defined for a specific context; we shall refer to such an approach as *external context checking*.

2. Check if the preconditions of the rule are satisfied; if yes — proceed; if no — go to the rule specified with the *else* part.
3. Retract all the facts that undergo the specification of the facts given in the *retract* part.
4. Assert all the facts given in the *assert* part.
5. Execute the actions defined with the conclusion part or present the appropriate decisions to the user.
6. Go to the rule specified with the *next* part.

In case the *next* or the *else* parts are empty, a default behavior can be assumed, e.g. the system can stop or go to examine next rule in the same table; if the current rule is the last one in the currently explored table, the system can either stop, try again beginning with the first rule in the table (e.g. monitor-like system) or backtrack and pass control to the former table. The details are domain-dependent and should be defined for any specific application.

For practical definition of certain details of operation, it should be stated precisely what it means to 'retract all the facts that undergo the specification of the facts given in the *retract* part'. The very first understanding may be that simply all the facts *identical* with the ones specified in the *retract* part of the rule should simply be removed. However, taking into account the precisely defined logical interpretation of the fact of the form

$$A_i = t_i$$

one may conclude that the retracting procedure should proceed as follows:

1. Examine all the facts specified in the *retract* part of the rule in turn.
2. For any such fact of the form $B_j = b_j$ determine all the facts in the knowledge base, defined with the use o the same attribute, i.e. of the form $A_i = a_i$, where $A_i = B_j$.
3. For any such fact check the following possibilities:
   - if $a_i \subseteq b_j$ then remove the fact $A_i = a_i$ from the knowledge base;
   - if $a_i \cap b_j \neq \emptyset$, then replace the fact $A_i = a_i$ in the database with $A_i = a_i \setminus b_j$;
   - if $a_i \cap b_j = \emptyset$, then leave the fact $A_i = a_i$ unchanged.

The system may have single-level construction, i.e. all the rules are considered to be of equal priority, or hierarchical — some of the rules may be considered to be of higher priority than the other. In particular, such system may be organized in several levels, where the rules at a higher level indicate which group of rules at a lower level should be used. The problems of inference control are discussed in Chap. 10.

8.5 Extended Tabular Trees

Extended Tabular Trees [92] (XTT, for short) combine the idea of extended attributive rule-based systems as presented in Chap. 8, Sects. 8.1–8.4, with the idea of decision trees as presented in Subsection 7.6.2. This allows for relatively high concentration of knowledge specification with extended attributive decision tables combined with the control capabilities incorporated in the structure of decision trees. In fact, the Extended Tabular Trees seems to constitute one of the most promising approaches to knowledge representation, fitting well the requirements for knowledge specification, analysis, verification and encoding control algorithms.

Let us present the anatomy of an XTT basic component for knowledge representation. Such a component is composed of a number of rules, defined as by (8.3), having the same structure of attributes and defined to operate in the same context. Such a set of rules can be efficiently represented with the use of a somewhat special table, being in fact an extension of (8.2).

In order to present the generic structure of an XTT knowledge representation component, let us introduce the following notions.

8.5.1 Cells

For intuition, a *cell* is the most atomic field for specification of information in tables. A cell including specific value constitutes a specification of atomic attributive formula of any of the introduced languages, i.e. an atom of AAL, SAL, VAAL, or VSAL.

More precisely, let A_j be an attribute of interest and let D_j denote its domain. Let also $i \in \{1, 2, \ldots, k\}$ denote some object or rule identifiers in some table T.

Definition 83. *A* cell *(in table T) is a pair*

$$(A_j, i).$$

A cell can be specified graphically — in this case it is represented as a rectangular field on the intersection of the column labelled with attribute A_j and the i-th row. A cell including a value — say t_{ij} — denotes in fact an atomic formula of the form

$$A_j(i) \in t_{ij}$$

if belonging to preconditions or

$$A_j(i) = t_{ij}$$

if belonging to conclusions.

Basic operations on a cell include writing a certain value down into a cell, removing the value, updating the value of a cell, and comparing the value of one cell with the value of another cell or an externally provided value.

8.5.2 Rules

Rules in tabular knowledge specification are equivalent to *rows* of the table. In fact, each rule of the form given by specification (8.3) can be represented with the use of a number of cells defining the preconditions and the output of the rule. A single rule is represented as a single row having the form presented in Table 8.5.

Table 8.5. The generic form of a rule in the XTT

Info		Prec	Retract	Assert	Decision	Ctrl	
I	Ctx	$A_1 \ldots A_n$	$B_1 \ldots B_b$	$C_1 \ldots C_c$	$H_1 \ldots H_h$	N	E
i	ψ	$t_{i1} \ldots t_{in}$	$b_{i1} \ldots b_{ib}$	$c_{i1} \ldots c_{ic}$	$h_{i1} \ldots h_{ih}$	g_i	e_i

Note that, it seems convenient to deal with the rule number, context, next and else specifications as values of certain specific attributes — this makes the specification of the table uniform. Hence, in the above single-row table, the columns are labeled with the following attributes:

- I — the rule identifier (in a table),
- Ctx — the context for the rule,
- A_1–A_n — the precondition attributes,
- B_1–B_b — the retract attributes,
- C_1–C_c — the assert attributes,
- H_1–H_h — the decision attributes,
- N — the control attribute indicating the *next* rule,
- E — the control attribute indicating the *else* rule.

The attribute may play a certain *role* — it can be used for specifying preconditions of the rule, retract or assert part of the rule or the output decision. Let

$$Roles = \{Info, Preconditions, Retract, Assert, Decision, Ctrl\}$$

denote the set for defining the six possible placements — and thus roles — of the attribute. Let also $i \in \{1, 2, \ldots, k\}$ denote rule identifier in a certain table T. We shall introduce the notation allowing to address any cell in the rule.

The idea of the notation follows the standard path notation used in relational databases. In fact, it is a specification of a path leading the cell and identifying it in a unique, systematic way. It is of the form

$$\langle role \rangle.\langle attribute \rangle,$$

where $\langle role \rangle \in \{Info, Preconditions, Retract, Assert, Decision, Ctrl\}$ and $\langle attribute \rangle \in \{I, Ctx, A_1 - A_n, B_1 - B_b, C_1 - C_c, H_1 - H_h, N, E\}$.

Further, to denote the specification of value of a certain cell in rule i we shall write

$$\langle role\rangle.\langle attribute\rangle(\langle rule\_identifier\rangle) = \langle value\rangle.$$

For example, $Info.I(i) = i$, $Info.Ctx(i) = \psi$ is a specification of rule context, $Preconditions.A_j(i) \in t_{ij}$ is a specification of precondition attribute restriction, $Ctrl.N(i) = g_i$ is a specification of the *next* rule, and $Ctrl.E(i) = e_i$ is a specification of the *else* rule.

8.5.3 XT — Extended Table

Rules of similar scheme can be easily combined into a table. The *Extended Attributive Table* (XAT) (or *Extended Table* (XT), for short) takes the following form (Table 8.6).

Table 8.6. The basic form of an XAT

Info		Prec	Retract	Assert	Decision	Ctrl	
I	Ctx	$A_1 \ldots A_n$	$B_1 \ldots B_b$	$C_1 \ldots C_c$	$H_1 \ldots H_h$	N	E
1	ψ	$t_{11} \ldots t_{1n}$	$b_{i1} \ldots b_{1b}$	$c_{11} \ldots c_{1c}$	$h_{11} \ldots h_{1h}$	g_1	e_1
2	ψ	$t_{21} \ldots t_{2n}$	$b_{21} \ldots b_{2b}$	$c_{21} \ldots c_{2c}$	$h_{21} \ldots h_{2h}$	g_2	e_2
$\vdots$	$\vdots$	$\vdots \ddots \vdots$	$\vdots \ddots \vdots$	$\vdots \ddots \vdots$	$\vdots \ddots \vdots$	$\vdots$	$\vdots$
i	ψ	$t_{i1} \ldots t_{in}$	$b_{i1} \ldots b_{ib}$	$c_{i1} \ldots c_{ic}$	$h_{i1} \ldots h_{ih}$	g_i	e_i
$\vdots$	$\vdots$	$\vdots \ddots \vdots$	$\vdots \ddots \vdots$	$\vdots \ddots \vdots$	$\vdots \ddots \vdots$	$\vdots$	$\vdots$
k	ψ	$t_{k1} \ldots t_{kn}$	$b_{k1} \ldots b_{kb}$	$c_{ik} \ldots c_{kc}$	$h_{k1} \ldots h_{kh}$	g_k	e_k

Note that the context specified by ψ is in fact the same for every rule in the table; it may be convenient to specify it outside the table and remove the *Ctx* column for simplicity.

In the most popular case also the values of the cells specifying the *next* and the *else* rule may also stay empty — this means that after examining a rule, perhaps the next one should be examined in turn, disregarding if the former rule was fired or not. If there is no next rule in the table, the interpreter is to stop and exit or perform other predefined or default actions.

The introduced above path notation can be extended onto tables in a straightforward way. The specification of a path is leading to the cell and identifying it in a unique, systematic way. It is of the form

$$\langle table\rangle.\langle role\rangle.\langle attribute\rangle,$$

where $\langle table\rangle$ is a name of a table, $\langle role\rangle \in \{Info, Preconditions, Retract, Assert, Decision, Ctrl\}$ and $\langle attribute\rangle \in \{I, Ctx, A_1 - A_n, B_1 - B_b, C_1 - C_c, H_1 - H_h, N, E\}$.

To denote the specification of value of a certain cell in table T we shall write

$$\langle table\rangle.\langle role\rangle.\langle attribute\rangle(\langle rule\_identifier\rangle) = \langle value\rangle.$$

For example, $T.Info.Ctx(i) = \psi$ is a specification of table and context for rule i, $T.Preconditions.A_j(i) \in t_{ij}$ is a specification of precondition attribute restriction, $T.Ctrl.N(i) = g_i$ is a specification of the *next* rule, and $T.Ctrl.E(i) = e_i$ is a specification of the *else* rule.

8.5.4 Connections and Their Properties

The extended tabular format provides an easy and intuitive way for specification of rules in tabular form. However, only similar rules operating in the same context can be put into a single table in a reasonable way. In more complex systems, there is no use of putting all rules to a single table — contrary to such a 'flat' approach, a number of tables with different structure can be specified.

In case of a single table rule-base, it seems that basic inference control paradigm can be specified in a relatively simple way. Typically, the rules are interpreted top-down, i.e. from the first one down until the last one is reached. Every time the preconditions of a rule are satisfied, the rule is fired. After a rule is examined, and possibly — fired, the next rule is analyzed. After reaching the last rule the process is stopped (in case of a single run) or it is continually repeated from the beginning (in case of a loop, to be applied in monitoring systems).

The basic inference scheme may be modified if the values of *Ctrl.N* and/or *Ctrl.E* cells are non-empty. A nonempty value of the *Ctrl.N* cell means that after successful rule firing, there is a jump to a specific rule. A nonempty value of the *Ctrl.E* cell means that after examining a rule which cannot be fired, there is a jump to a specific rule. This allows to specify almost any control algorithm.

Having a number of such tables the *Ctrl.N* and/or *Ctrl.E* cells can be used to specify *switching among tables* with the possibility to jump into a specific rule inside a table. Hence, with the use of the control cells and assuming that rules in a table are numbered, one can easily specify links between tables.

In order to specify a link, the following three elements are necessary:

1) starting point — in our case it is the value of the location of the *Ctrl.N* and *Ctrl.E* cell,
2) end point — in our case it is a rule in a goal table,
3) definition of properties (if any) of the link.

In order to identify a rule in a unique way we adopt the path-dot notation; a path $\langle table\rangle.\langle number\rangle$ indicates a precise rule located in a table. Hence, in order to specify the values of g_i and e_i (the *next* and the *else* rules) one should use the name (identifier) of the goal table and the number of the rule inside it. So, for example, $T.i$ is the identifier of rule i in table T.

As for the properties, there is one important characteristic to be taken into account. This is the *cut* property, analogous to the *cut* predicate in PROLOG, see Subsect. 11.6.1 on page 182.

Assume that the control has been passed from table S, rule i to table T, rule j. Then, there is an attempt to check the preconditions of rule $T.j$ and fire it, which can be successful or not. Now, assume that there is no rule to be executed after $T.j$, perhaps because the cells *Ctrl.N* and *Ctrl.E* of the rule are empty and there is no other rule in table T. Should the inference control mechanism stop and exit in such a case?

Basically, there are two possibilities. The link $(S.i, T.j)$ can allow for backtracking, and we may assume that the system backtracks to table S and resumes there, trying the next rules after rule $S.i$. Assume this is the default mode of operation. The second possibility is to forbid backtracking, and then the system is to stop operation. In the second case we would require an explicit specification of the *cut* as a property of the link. The cut is denoted with ! as in the PROLOG language. Hence, $(S.i, T.j)$ specifies a link allowing for backtracking (default), while $(S.i, !, T.j)$ (or $(S.i!, T.j)$) specifies a link with no-backtracking property. For simplicity, the ! symbol may be appended directly to the values of the *Ctrl.N* and *Ctrl.E* cells.

8.6 Example: Thermostat

Let us consider a simple but illustrative rule-based control system for setting the required temperature in a room, depending on the type of the day, season, hours, etc. The example is based on [102] THERMOSTAT example[7].

```
/* THERMOSTAT: A DEMONSTRATION RULE-BASE */
Rule: 1
    if      the day is Monday
    or      the day is Tuesday
    or      the day is Wednesday
    or      the day is Thursday
    or      the day is Friday
    then    today is a workday
Rule: 2
    if     the day is Saturday
    or     the day is Sunday
    then    today is the weekend
Rule: 3
    if      today is workday
```

[7] An appropriate knowledge base is given in [102], pages 41–43; for convenience we list it here once again. Note that, from the point of view of *control theory* the specification is not one of a *thermostat* system (which has as a task temperature stabilization at a certain *set point*, but it is a specification of *set point* selection algorithm of a higher-level (adaptation) controller.

```
    and   the time is ‘between 9 am and 5 pm’
    then  operation is ‘during business hours’
Rule: 4
    if    today is workday
    and   the time is ‘before 9 am’
    then  operation is ‘not during business hours’
Rule: 5
    if    today is workday
    and   the time is ‘after 5 pm’
    then  operation is ‘not during business hours’
Rule: 6
    if    today is weekend
    then  operation is ‘not during business hours’
Rule: 7
    if    the month is January
    or    the month is February
    or    the month is December
    then    the season is summer
Rule: 8
    if    the month is March
    or    the month is April
    or    the month is May
    then    the season is autumn
Rule: 9
    if    the month is June
    or    the month is July
    or    the month is August
    then    the season is winter
Rule: 10
    if    the month is September
    or    the month is October
    or    the month is November
    then    the season is spring
Rule: 11
    if    the season is spring
    and   operation is ‘during business hours’
    then  thermostat_setting is ‘20 degrees’
Rule: 12
    if    the season is spring
    and   operation is ‘not during business hours’
    then  thermostat_setting is ‘15 degrees’
Rule: 13
    if    the season is summer
    and   operation is ‘during business hours’
    then  thermostat_setting is ‘24 degrees’
Rule: 14
    if    the season is summer
    and   operation is ‘not during business hours’
    then  thermostat_setting is ‘27 degrees’
```

```
Rule: 15
    if    the season is autumn
    and   operation is 'during business hours'
    then  thermostat_setting is '20 degrees'
Rule: 16
    if    the season is autumn
    and   operation is 'not during business hours'
    then  thermostat_setting is '16 degrees'
Rule: 17
    if    the season is winter
    and   operation is 'during business hours'
    then  thermostat_setting is '18 degrees'
Rule: 18
    if    the season is winter
    and   operation is 'not during business hours'
    then  thermostat_setting is '14 degrees'
```

Note that the presented rules can be divided in a natural way into groups producing the same kind of decision, but the precise decision depends on precise preconditions. Further, in the groups, the preconditions of the rules employ the same linguistic variables, but different values of them. In fact, there are four different groups of rules defining decisions whether today is workday or weekend (rules 1 and 2), whether the time is during business hours or not (rules 3–6), what season we have (rules 7–10), and finally indicating the setting of the thermostat (rules 11–18).

For each group we can build a separate extended tabular system — this seems to be a natural and efficient approach — since the rules use the same variables, the specification of attributes in the table is just given once and all the columns are necessary for any rule in a table. Further, using the *next* and *else* specification of rules we can specify an efficient way of interpretation of the rules.

Below, we shall provide specification of this rule-base using the XTT approach. We provide specification of the XT tables and connections among them. For simplicity, the context for each table will be provided once, outside of the table.

In order to make the tables concise we shall introduce the following shorthand notation for attributes and their values.

Here is the list of attributes to be considered; the name of each attribute starts with 'a':

- *aDD* — day,
- *aTD* — today,
- *aTM* — time,
- *aOP* — operation,
- *aMO* — month,
- *aSE* — season,
- *aTHS* — thermostat_setting.

In the specification, the following values of the attributes will be used; the name of each set starts with 's':

- $sWD = \{Monday,\ Tuesday,\ Wednesday,\ Thursday,\ Friday\}$,
- $sWK = \{Saturday,\ Sunday\}$,
- $sSUM = \{January,\ February,\ March\}$,
- $sAUT = \{March,\ April,\ May\}$,
- $sWIN = \{June,\ July,\ August\}$,
- $sSPR = \{September,\ October,\ November\}$,
- sum = 'summer',
- aut = 'autumn',
- win = 'winter',
- spr = 'spring',
- wd = 'workday',
- wk = 'weekend'
- dbh = 'during business hours',
- $ndbh$ = 'not during business hours'.

Note that we have introduced set values for attributes such as aDD (day) and aSE (season) — this allows to specify the so-called *internal disjunction* in a very concise way.

For simple interpretation and analysis, the original numbering of rules is kept over the tables.

For the rules 1 and 2 we have:

Table 8.7. Context 1: none. An XT for rules 1 nad 2

Info	Prec	Retract	Assert	Decision	Ctrl	
I	aDD	aTD	aTD	H	N	E
1	sWD	—	wd		2.3	1.2
2	sWK	—	wk		2.6	1.1

Specification of the retract part with the underscore means in fact that any value of aTD should be removed (the variables should be cleared). The same convention is applied in further tables.

Note also the way of specifying the *else* part of the rules — the provided specification assures circular attempt at interpretation of the rules until the value of aTD attribute is set. This is so since this value is necessary in further inference. Once the value is successfully established, the control is passed to the tables indicated in the *next* part. The specification of the *next* rules is made in such a way that depending on the value of aTD it is passed to precisely selected rules having a chance to be fired.

For rules 3, 4, 5, and 6 we have:

Table 8.8. Context 2: $aTD \in \{wd, wk\}$. An XT for rules 3–6

Info	Prec		Retract	Assert	Decision	Ctrl	
I	aTD	aTM	aOP	aOP	H	N	E
3	*wd*	[9:00, 17:00]	—	*dbh*		3.7	2.4
4	*wd*	[00:00, 09:00]	—	*ndbh*		3.7	2.5
5	*wd*	[17:00, 24:00]	—	*ndbh*		3.7	2.6
6	*wk*	—	—	*ndbh*		3.7	2.3

As before, note the way of specifying the *else* part of the rules — the provided specification assures circular attempt at interpretation of the rules until the value of aOP attribute is set. This is so since this value is necessary in further inference. Once the value is successfully established, the control is passed to the table indicated in the *next* part. The specification of the *next* rules is the same over the whole column of Table 8.8, since in Table 8.9 we have to start at the beginning.

For rules 7, 8, 9, and 10 we have:

Table 8.9. Context 3: none. An XT for rules 7–10

Info	Prec	Retract	Assert	Decision	Ctrl	
I	aMO	aSE	aSE	H	N	E
7	*sSUM*	—	*sum*		4.13	3.8
8	*sAUT*	—	*aut*		4.15	3.9
9	*sWIN*	—	*win*		4.17	3.10
10	*sSPR*	—	*spr*		4.11	3.7

For the final decision rules 11, 12, 13, 14, 15, 16, 17, and 18 we have:

Table 8.10. Context 4: $aSE \in \{spr, sum, aut, win\} \wedge aOP \in \{dbh, ndbh\}$. An XT for rules 11–18

Info	Prec		Retract	Assert	Decision	Ctrl	
I	aSE	aOP			$aTHS$	N	E
11	*spr*	*dbh*			20	1.1	4.12
12	*spr*	*ndbh*			15	1.1	4.13
13	*sum*	*dbh*			24	1.1	4.14
14	*sum*	*ndbh*			17	1.1	4.15
15	*aut*	*dbh*			20	1.1	4.16
16	*aut*	*ndbh*			16	1.1	4.17
17	*win*	*dbh*			18	1.1	4.18
18	*win*	*ndbh*			14	1.1	1.1

In the tables 8.7–8.9 the output values are asserted to the global memory since they are used during inference with the Table 8.10. The final, output conclusion is produced with the rules of the Table 8.10 — it is materialized as a setting of the working point for the thermostat system.

The system can be run occasionally, e.g. it can be activated after a specific event such as passing through a specific value of time requiring changing the settings, or it can work in the monitoring mode, repeating continually execution of the rules. Here we have selected the second option, and thus after successful setting of the desired thermostat temperature the control is passed to the beginning; the same happens if no rule of the Table 8.10 is fired. This period of evaluation of the rules may be quite long, e.g. it may take one hour.

Finally, one can ask for explicit specification of links among the tables. The specification follows from the declaration of the *next* and the *else* parts of the rules. The declaration of the inter-tabular links is as follows:

$(N, 1.1, !, 2.3)$ – after successful firing of rule 1 in Table 8.7 the control is passed to rule 3 in Table 8.8; the ! means that there is no use in backtracking (in fact, preconditions of rules 1 and 2 are mutually exclusive);

$(N, 1.2, !, 2.6)$ – after successful firing of rule 2 in Table 8.7 the control is passed to rule 6 in Table 8.8; the ! means that there is no use in backtracking;

$(N, 2.3, !, 3.7)$ – after successful firing of rule 3 in Table 8.8 the control is passed to rule 7 in Table 8.9; the ! means that there is no use in backtracking;

$(N, 2.4, !, 3.7)$ – after successful firing of rule 3 in Table 8.8 the control is passed to rule 7 in Table 8.9; the ! means that there is no use in backtracking;

$(N, 2.5, !, 3.7)$ – after successful firing of rule 3 in Table 8.8 the control is passed to rule 7 in Table 8.9; the ! means that there is no use in backtracking;

$(N, 2.6, !, 3.7)$ – after successful firing of rule 3 in Table 8.8 the control is passed to rule 7 in Table 8.9; the ! means that there is no use in backtracking;

$(N, 3.7, !, 4.13)$ – after successful firing of rule 7 in Table 8.9 the control is passed to rule 13 in Table 8.10; the ! means that there is no use in backtracking;

$(N, 3.8, !, 4.15)$ – after successful firing of rule 8 in Table 8.9 the control is passed to rule 15 in Table 8.10; the ! means that there is no use in backtracking;

$(N, 3.9, !, 4.17)$ – after successful firing of rule 9 in Table 8.9 the control is passed to rule 17 in Table 8.10; the ! means that there is no use in backtracking;

$(N, 3.10, !, 4.11)$ – after successful firing of rule 10 in Table 8.9 the control is passed to rule 11 in Table 8.10; the ! means that there is no use in backtracking;

$(N, 4.11, !, 1.1)$ – after successful firing of rule 11 in Table 8.10 the control is passed to rule 1 in Table 8.7; the ! means that there is no use in backtracking;

$(N, 4.12, !, 1.1)$ – after successful firing of rule 12 in Table 8.10 the control is passed to rule 1 in Table 8.7; the ! means that there is no use in backtracking;

$(N, 4.13, !, 1.1)$ – after successful firing of rule 13 in Table 8.10 the control is passed to rule 1 in Table 8.7; the ! means that there is no use in backtracking;

$(N, 4.14, !, 1.1)$ – after successful firing of rule 14 in Table 8.10 the control is passed to rule 1 in Table 8.7; the ! means that there is no use in backtracking;

$(N, 4.15, !, 1.1)$ – after successful firing of rule 15 in Table 8.10 the control is passed to rule 1 in Table 8.7; the ! means that there is no use in backtracking;

$(N, 4.16, !, 1.1)$ – after successful firing of rule 16 in Table 8.10 the control is passed to rule 1 in Table 8.7; the ! means that there is no use in backtracking;

$(N, 4.17, !, 1.1)$ – after successful firing of rule 17 in Table 8.10 the control is passed to rule 1 in Table 8.7; the ! means that there is no use in backtracking;

$(N, 4.18, !, 1.1)$ – after successful firing of rule 18 in Table 8.10 the control is passed to rule 1 in Table 8.7; the ! means that there is no use in backtracking;

$(E, 4.18, !, 1.1)$ – after checking that firing of rule 18 in Table 8.10 is impossible the control is passed to rule 1 in Table 8.7; the ! means that there is no use in backtracking.

The specification of the system with the use of the extended decision tables seems to be both concise and easy to analysis; we return to verification and design problems of such systems in further parts of the book.

9

Rule-Based Systems in First-Order Logic

First-Order Logic constitutes a really powerful language for specification of rule-based systems. This is thanks to the expressive power of First-Order Predicate Calculus (FOPC), including the possibility of using variables, structural terms, and predicates of arbitrary arity. Further, negation can be used in explicit or implicit manner. Moreover, in definition of rules some meta-language features, such as the concepts of *retract* and *assert* predicates and inference control elements can also be used.

9.1 Basic Form of Rules

Let Φ and Ψ be any First Order Logic formulae of arbitrary complexity. In fact, some most general form of rules can be as follows

$$\text{rule}\colon \Phi \longrightarrow \Psi. \tag{9.1}$$

Such a basic scheme can be used for expressing quite complex knowledge in form of rules; however, for practical applications, and especially keeping in mind the simplicity required for automated inference, it is assumed that the basic form will be analogous to the one of propositional logic or attributive logic rules.

Let $q_1, q_2, \ldots, q_n$ and h be some literals of First Order Logic. The most typical form of a rules is as follows

$$q_1 \wedge q_2 \wedge \ldots q_n \longrightarrow h. \tag{9.2}$$

In the above rule it is assumed that individual atomic formulae used for defining precondition formula may be individually negated (if necessary) and that the only other logical connective used for defining precondition formula is the conjunction. No quantifiers are used explicitly. The conclusion is composed of a single literal, mostly positive one. Variables — if any — are assumed to be universally quantified. Further, it is usually assumed that there is

A. Ligêza: *Logical Foundations for Rule-Based Systems*, Studies in Computational Intelligence (SCI) **11**, 155–162 (2006)
www.springerlink.com

$FV(h) \subseteq FV(q_1) \cup FV(q_2) \cup \ldots FV(q_n)$, i.e. having established the variables in the preconditions, all the variables of conclusion are also defined.

9.2 FOPC Rule-Base Example: Thermostat

As for example, let us consider one of the possible specifications of the thermostat example (see Sect. 8.6) as a First-Order Logic rule-base. For simplicity, we shall use the names of the attributes as single argument predicates.

The specification p/k means that predicate p has k arguments. We have the following predicates with assigned interpretations:

- $aDD/1$ — the day is,
- $aTD/1$ — today is wd (workday) or wk (weekend),
- $aTM/1$ — the time is; we use $int(a, b)$ to say that the time is within the interval determined by $[a, b]$,
- $aOP/1$ — operation is dbh (during business hours) or $ndbh$ (not during business hours),
- $aMO/1$ — the month is,
- $aSE/1$ — the season is sum (summer), aut (autumn), win (winter) or spr (spring),
- $aTHS/1$ — the thermostat setting is.

For defining sets we use the notation of lists; hence, instead of using constructs such as $l(saturday, l(sunday, nil))$ we simply put $[saturday, sunday]$ (see rule 2).

We have the following rules.

Rule 1: $aDD([monday, tuesday, wednesday, thursday, friday]) \longrightarrow aTD(wd)$.
Rule 2: $aDD([saturday, sunday]) \longrightarrow aTD(wk)$.
Rule 3: $aTD(wd) \wedge aTM(int(9, 17)) \longrightarrow aOP(dbh)$.
Rule 4: $aTD(wd) \wedge aTM(int(0, 8)) \longrightarrow aOP(ndbh)$.
Rule 5: $aTD(wd) \wedge aTM(int(18, 24)) \longrightarrow aOP(ndbh)$.
Rule 6: $aTD(wk) \longrightarrow aOP(ndbh)$.
Rule 7: $aMO([january, february, december]) \longrightarrow aSE(sum)$.
Rule 8: $aMO([march, april, may]) \longrightarrow aSE(aut)$.
Rule 9: $aMO([june, july, august]) \longrightarrow aSE(win)$.
Rule 10: $aMO([september, october, november]) \longrightarrow aSE(spr)$.
Rule 11: $aSE(spr) \wedge aOP(dbh) \longrightarrow aTHS(20)$.
Rule 12: $aSE(spr) \wedge aOP(ndbh) \longrightarrow aTHS(15)$.
Rule 13: $aSE(sum) \wedge aOP(dbh) \longrightarrow aTHS(24)$.
Rule 14: $aSE(sum) \wedge aOP(ndbh) \longrightarrow aTHS(17)$.
Rule 15: $aSE(aut) \wedge aOP(dbh) \longrightarrow aTHS(20)$.
Rule 16: $aSE(aut) \wedge aOP(ndbh) \longrightarrow aTHS(16)$.
Rule 17: $aSE(win) \wedge aOP(dbh) \longrightarrow aTHS(18)$.
Rule 18: $aSE(win) \wedge aOP(ndbh) \longrightarrow aTHS(14)$.

Note that instead of using implicit negation as in rule 4, we can also write

$$\text{rule 4'}: \; aTD(wd) \wedge aTM(int(0,8)) \longrightarrow \neg aOP(dbh).$$

Negation can also be used in preconditions; for example, instead of using *ndbh* in rule 18 we could have

$$\text{rule 18'}: \; aSE(win) \wedge \neg aOP(dbh) \longrightarrow aTHS(14).$$

In a more advanced formulation, instead of rules 4 and 5 one could perhaps use a single rule of the form

$$\text{rule } (4+5): \; aTD(wd) \wedge \neg aTM(int(9,17)) \longrightarrow \neg aOP(dbh).$$

The advantages of using First-Order Logic are visible — the notation becomes concise, and the expressive power is high. Moreover, readability of the code is close to natural language. The code becomes short, since unnecessary conditions in preconditions can just be omitted. On the other hand, processing full first order rules becomes a more complex task, and analysis of them may become much more complex than in the case of attributive languages.

In the next section it is shown that the basic syntactic form of rules can become much more complex, including specification of dynamic modification of the fact base and control features.

9.3 Extended Form of FOPC Rules

In an extended, generalized form any of rules should contain several parts, i.e. a part defining when it is possible to apply the particular rule (*preconditions*), a part specifying the action to be taken (*action*), and a part specifying the changes in the state description. The changes of current state formula can be executed by retracting the facts which after the action specified by the rule are no longer true (*delete_results*) and the ones which become true as a result of the action (*add_results*). The idea of the transformation rules is based on the STRIPS[1] representation of operators (see [31, 104]).

The basic format for any dynamic rule considered in this Chapter is as follows:

```
if          precondition(s)
then
      do        action(s)
      retract   delete_result(s)
      assert    add_result(s)
```

[1] STRIPS stands for Stanford Research Institute Problem Solver; it was one of the first rule-based planning systems; see [31].

The interpretation of this scheme is straightforward: if *precondition(s)* is (are) true (i.e. satisfied with regard to the current state), then the specified *action(s)* is (are) to be executed, the knowledge base constituting the current state formula is to be modified (updated) by retracting *delete_result(s)* (the ones which are no longer true) and asserting *add_result(s)* (the ones which become true).

In the presented scheme of rules the *preconditions* are specified with the use of a First Order Logic formula; for simplicity, in most cases it is a simple conjunctive formula (or a formula in DNF form). The *preconditions* specify in fact a joint description of all the states for which the specified rule is applicable. The *delete_results* and *add_results* are just sets (lists) of First Order Logic facts.

Any of the rule components can contain parameters, i.e. variables. Since the formulation of rules is normally quantifier-free, a note on quantification of these variables may be in order. In general, from logical point of view one can consider all variables in a rule to be *universally* quantified, similarly as in the case of Prolog clauses. From practical point of view the variables denote *parameters* to be instantiated before each application of a rule. This means in fact that any rule with variables denotes *a set* of (or a general scheme of) a number of applicable rules, each of them to be obtained from the general scheme by appropriate substituting some terms for the variables — in this way a concretization of the rule for execution is obtained.

The precise values of parameters may be selected by a user (or any domain-specific selection mechanism) and then the preconditions are to be proved, or — and this is a more frequent approach — the values of variables for which preconditions are satisfied are found during the process of proving preconditions (as a side effect of unification and factorization).

On the other hand, in the case of checking if the precondition formula covers the current state formula, all the variables of preconditions may be assumed to be *existentially* quantified, just for 'adjusting' the rule to the current state and purposes. In order to fire a rule it is enough to find a single instantiation of the variables so that preconditions are satisfied. The values of the variables can be found by a rule matching mechanism, i.e. during the process of verification if the rule can be applied, or they can be given by meta-control mechanisms, e.g. a goal defining one or by the user.

For intuition, consider the so-called Block World example. There are some blocks, say three, which can be placed on the floor or on one another. Assume we have blocks labelled `a`, `b` and `c`. Some example states are presented in Fig. 9.1.

Let us specify rules describing potential manipulations of these blocks. A block can be moved from its location to another location. A location for a block can be the floor (with unlimited space) or the top surface of another block. Only one block can be moved at a time. Below, we specify three rules describing possible state changes. These transformation rules define three possible actions which can be undertaken in the system.

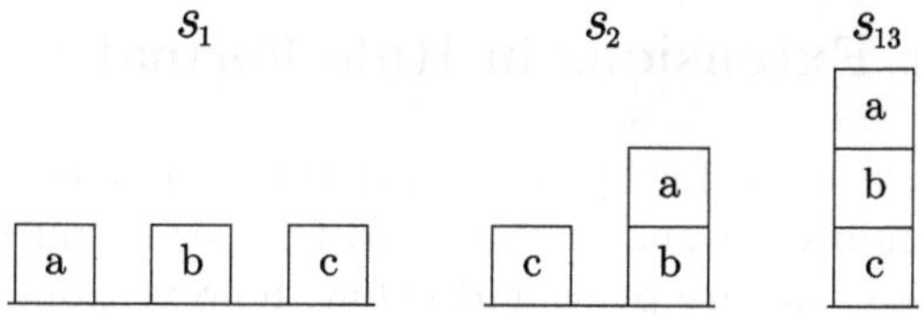

Fig. 9.1. Example states of the Block World

```
rule(1):  clear(Y)
if                  on(X,Y) ∧ cleartop(X)
then
          do        move X to the floor
          retract   on(X,Y)
          assert    onfloor(X), cleartop(Y)
rule(2):  puton(X,Y)
if                  onfloor(X) ∧ cleartop(X) ∧ cleartop(Y)
then
          do        move X from the floor onto Y
          retract   onfloor(X),cleartop(Y)
          assert    on(X,Y)
rule(3):  move(X,Y,Z)
if                  on(X,Y) ∧ cleartop(X) ∧ cleartop(Z)
then
          do        move X from Y onto Z
          retract   on(X,Y),cleartop(Z)
          assert    on(X,Z), cleartop(Y)
```

One can see that these three operations — namely `clear(X)`, `puton(X,Y)`, and `move(X,Y,Z)` — specify all the possible state changes, i.e. taking a block from another one and putting it on the floor, taking a block from the floor and putting it on the top of another one, and moving a block from some other block onto a third one. By an appropriate sequence of these actions (with appropriate parameters instantiation) one can achieve transformation of any initial state of the system into a desired goal state.

Note that the scheme subsumes both the STRIPS format ([31, 104]), and the typical formats of rules used in expert systems ([46, 137]). In fact, it is powerful enough for encoding even complex operations. However, for certain specific purposes the above basic format may be insufficient. Thus, various possible modifications and extensions can be applied; some most typical ones are presented in the next section.

9.4 Further Extensions in Rule Format

The given above basic format of transformation rules can be adjusted and extended depending on and with regard to specific purposes. The most important modifications are presented below; however, one can introduce many further elements which one finds necessary.

Firstly, rules can be usually grouped into some sets of them, designed to be operating in a specific context. It may be convenient to specify the *context formula* as an additional, excluded part of the information when the rule is likely to be applied. Note, that with regard to the possibility of specifying in *preconditions* any arbitrarily complex formula, the above modification can be considered to introduce an artificial split of information covered by the precondition formula; but in practice explicit definition of context can be very useful, both for improving efficiency of rule execution and for analysis, verification and design. This is so since when operating in certain context only limited set of rules is to be further analyzed. The modification can consist of adding a part of the form

if *context_formula*

which itself can have more than one occurrence in the rule, e.g. for subcontexts, etc.

Secondly, for some rules and in certain systems, it may be convenient to specify as an additional part the information when the rule certainly *cannot be used*. Note, that with regard to the possibility of specifying in *preconditions* any arbitrarily complex formula, the above modification can be considered to introduce a redundant information. However, in certain cases it may be profitable to prune (reject) some or most of the rules in a simple and efficient manner, since finding that a rule is inapplicable can be sometimes much faster. This is the case of rarely used rules, the use of which can be excluded by a simple condition, e.g. a single fact. The modification can consist of adding a part of the form

if not *excluding_condition(s)*

which itself can have more than one occurrence in the rule. Further, let us notice that the above modification can be included before the *preconditions* — as for quick pruning of the rule in case of inapplicability, or after *preconditions* — for some further refinement, when some of the parameters have already been established during preconditions verification.

Thirdly, in certain systems it may be the case, that after application of some rule operating in certain context it is very likely (or even sure) that the next rule to be applied is determined in a unique way, or a subset of the set of rules to be applied next can be given. In the first case the rule directly points out to some other rule and we call it *rule switching*; in the latter case the rule points to some set of rules and we call it *context switching*. It may be reasonable to specify this rule (these rules) explicitly, so as to enable the reasoning control mechanism to change the order of rule examination. The rule interpretation mechanism is discussed later on. Now let us present the additional part of the rules which looks as follows

next *rule(s).*

Such a specification can significantly speed up the execution of rules, and thus improve the overall efficiency of the system.

It may be also the case that the next rule (rules) is (are) determined in the case the current rule fails to be fired; the appropriate specification can take the form

else *rule(s).*

In the above specifications the pointer to rule or rules can be implemented through specification of rule number, unique rule name, or name of a set of rules and a single rule to start within this set, as in the case of tabular knowledge specification.

Fourthly, it may be useful to specify the resources necessary to execute the action described by a rule. This may happen for example in case of *parallel rule execution* — some rules may require exclusive use of certain resources (they may block the use of certain resources by other rules). The information may be necessary for the reasoning control mechanism in order to resolve some potential conflicts. The possible modification may look as follows

resources *resource(s).*

The simplest way of specifying the *resources* is to provide a list (set) of them, however, this part may also contain specialized routines for checking if a conflict with regard to resources occurs, etc.

Finally, in monitoring systems with a human supervisor it may be useful to specify an additional part containing a message for the operator. Thus some of the rules may be equipped with the following part

output *message(s).*

The above information, sent to the console or an archive file, can be parameterized, i.e. contains knowledge of current process parameters.

After combining together the presented above possibilities and adding a rule number to any of the rules, the following potential rule scheme is obtained:

<u>**rule**(n)</u> *name(parameters)*
resources *resource(s)*
if *context_formula*
and
if not *excluding_condition(s)*
and
if *precondition(s)*

	and	
if not		*detailed_excluding_condition(s)*
then		
	do	*action(s)*
	retract	*delete_result(s)*
	assert	*add_result(s)*
	output	*message(s)*
next		*rule(s)*
else		*rule(s)*

with the given above interpretation of the components.

In the next section a discussion of inference control problems is provided and a way of executing such complex rules is outlined.

10

Inference Control in Rule-Based Systems

The chapter is devoted to presentation of the issues concerning rule interpretation, inference control and conflict resolution strategies in rule-based systems. A knowledge-based system is considered to be one composed of a set of complex, frame-like rules, each of them having internal structure and different components. Any of such rules can be regarded as a specialized procedure being able to perform a specific task it is designed for, provided that it is activated in a situation satisfying its preconditions. The problem of conflict resolution arises in non-deterministic systems, when two or more rules can be activated at a time.

The problem of conflict resolution is outlined and a review of conflict resolution strategies is provided. The conflict resolution problem is stated in its basic and advanced form. In the basic form the solution is accomplished by selection of a single rule to be executed. The discussed strategies include ones based on linear order, linear order with immediate repetition from the beginning, and priority-based ordering. The advanced form of the conflict resolution problem statement includes selection of the conflict set and rule/rules selection and firing with possible passing the control to another mechanisms.

The discussion of reviewed strategies include several classifications, e.g. into static vs. dynamic strategies, syntactic vs. semantic ones, direct vs. indirect, based on simple criteria, modifiable/adaptable and learning ones. The strategies reviewed range from the simplest ones based on syntactic features of the rules, through ones taking into account the context, various priority evaluation mechanisms and adaptable features to strategies based on meta-inference systems and learning. Some criteria determining intelligent conflict resolution are put forward.

The main contribution of this chapter consists of providing a new conflict resolution mechanism for rule-based systems including real-time control systems. The mechanism is based on the specific, frame-like structure of the rules. This structure includes the most important part for control, i.e. built-in-rule context-sensitive inference control part.

A. Ligêza: *Logical Foundations for Rule-Based Systems*, Studies in Computational Intelligence (SCI) **11**, 163–171 (2006)
www.springerlink.com

The overall rule-selection strategy is two-fold: firstly, there exists an external, global rule interpreter responsible for selection of the conflict set and conflict resolution (including matching the preconditions of the rules and firing them); secondly, depending on the context, the control can be passed to a lower-level mechanism encoded directly in a rule.

An auxiliary control mechanism is provided through the assert/retract parts of rules playing the role of a blackboard-type communication mechanism. It is pointed out that the mechanism is flexible enough to imitate most of the presented classical strategies, and simultaneously can constitute an efficient tool for on-line conflict resolution provided that the right cooperation is ensured by expert control strategy.

10.1 Problem Statement

Consider a dynamic system composed of n rules. Conflict resolution problem can be stated as follows: given a set of n rules, select a single rule to be applied in the current state of the system under control/supervision.

The conflict resolution problem does not exist in *deterministic* systems [53], i.e. ones for which only one rule can be applied in any state. The problem occurs if more than one rule can be applied for a state, and, in most of the systems a single rule is selected with respect to some auxiliary criteria, different from the applicability ones. Depending on the complexity of the structure of inference control mechanism two basic formulations, a basic and an advanced one can be put forward.

10.1.1 Basic Problem Formulation

Consider a rule-based control system [53] composed of n rules $r_1, r_2, \ldots, r_n$. Assume that the rules are ordered linearly according to their indices. The *conflict resolution problem* is defined as the task of selecting exactly one rule applicable in the current state. The selected rule should be applied and the process is repeated from the beginning.

The basic strategies for approaching the above problem may vary depending on system specification and application. There are, however, two basic possibilities:

1) a rule r_i appropriate for achieving a currently desired goal can be selected first; then an attempt to satisfy its preconditions can be undertaken; this approach is mostly applied in planning systems;
2) the rules are sequentially tested for satisfaction of their preconditions, and the first one with satisfied preconditions (and appropriate for the specific task performed, if some additional requirements are specified) is selected and executed.

With respect to control and supervision problems mostly the second approach is applied.

There are four basic strategies for approaching the problem:

1) **Closed-loop hierarchical linear strategy:** a hierarchy amongst the rules is assumed. The rules are tested sequentially until one with satisfied preconditions is found. Then the rule is executed and the process is resumed from rule 1. In PROLOG this would correspond to the *run-find-execute-!-run* loop scheme.

2) **Closed-loop linear strategy:** the rules are ordered linearly and satisfiability of their preconditions is tested sequentially in a closed loop. After rule i rule number $i+1$ is tested and possibly executed; after rule n rule 1 is taken into account. In PROLOG this scheme would correspond to the *repeat-find-execute-fail* loop format.

3) **Linear strategy with rule-switching:** linear interpretation from the beginning to first executable rule (one with satisfied preconditions); then the system jumps to a rule or rules indicated by the recently executed rule. This kind of scheme consists of switching among rules. This approach corresponds to the use of the *next* part in rules.

4) **Linear strategy with context-switching:** finally, linear rule examination may be combined with context switching, i.e. after a rule is executed, the current situation of the system is evaluated, the current context (qualitative situation) is determined and a decision making mechanism switches to the set of rules appropriate for this context.

Finally, in more complex systems a subset of the set of initial rules may be selected (a set of rules for a specific context) and re-ordered according to the priorities (stable or depending on current context). The rules are then inspected with respect to any of the two above schemes.

The types of interpreters are schematically presented in the pictures (Figs. 10.1 and 10.2).

Of course, the above approaches can be modified and extended with the use of auxiliary reasoning control mechanisms, rule preselection tools, etc., admitting open-loop only, using *cut* operator to prohibit backtracking, etc.

10.1.2 Advanced Problem Formulation

The advanced form of the conflict resolution problem statement includes selection of the conflict set and rule/rules selection and firing with possible passing the control to another mechanism. Again, consider a system composed of n rules, say $R = \{r_1, r_2, \ldots, r_n\}$.

The advanced formulation of the conflict resolution task is as follows: select a subset $R' \subseteq R$ of rules with satisfied preconditions, and if there are two or more such rules select one satisfying some auxiliary criteria.

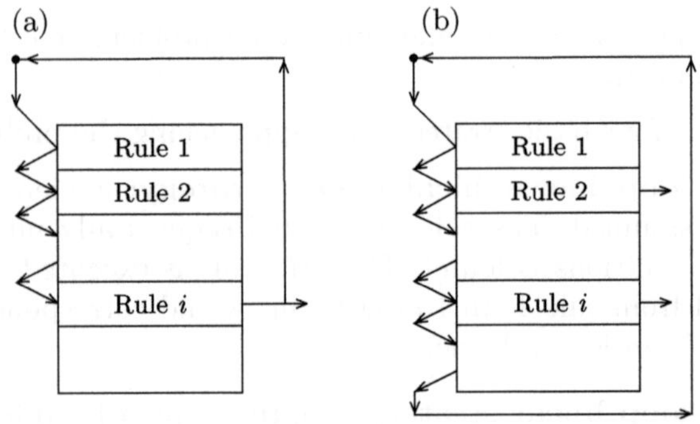

Fig. 10.1. Linear interpreters: a) Closed-loop hierarchical linear strategy; b) Closed-loop linear strategy

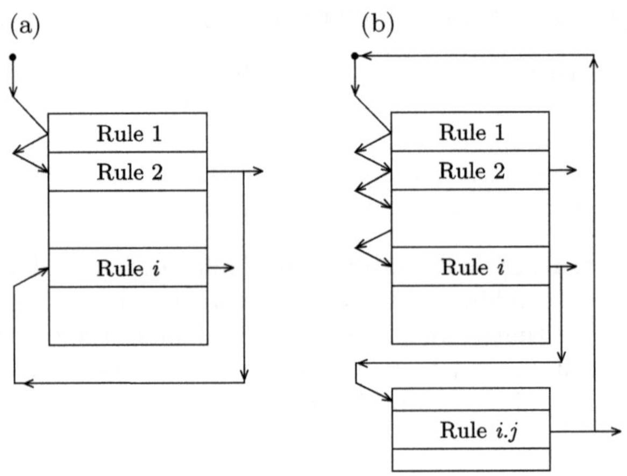

Fig. 10.2. Switching interpreters: (**a**) Rule switching; (**b**) Context switching

Thus the approach consists of two stages:

1) initial elimination of rules which cannot be applied,
2) selection of a rule which is best, appropriate or likely for performing the current control task.

For the advanced case several classifications of conflict resolution strategies can be put forward:

- static vs. dynamic strategies; static strategies are based on criteria constant over time, while dynamic ones can take into account current context, time, number of (successful) repetitions of a rule, etc.;
- syntactic vs. semantic strategies; the first one base on the 'shape' of the rule preconditions, while the second ones may take into account the current context, desired goal, and evaluable user-specified criteria;

- direct vs. indirect strategies; the direct ones are based on simple comparison of rules and 'ordering factors' assigned to them, e.g. priorities, while the indirect strategies can be implemented with an auxiliary knowledge-based system, meta-rules and complex inference schemes;
- strategies based on simple, constant criteria vs. ones modifiable/adaptable and learning.

Depending on the problem, its specification and resources available various strategies can be applied; no general criteria can be put forward. Compositions of several strategies may also be useful.

10.2 Rule Interpretation Algorithm

Taking into account the basic formulation of the problem, the important component of the strategy is the one for interpretation of a single selected rule. A first rule to be executed can be defined in an arbitrary way or the set of rules can be searched according to some order. The next rules can be indicated by the *next* and *else* components. The algorithm for interpretation of a single selected rule will be specified below.

Let ϕ be the state formula (the current knowledge base concerning state of the system).

Below, a symbolic representation of components of the extended rule format is introduced:

rule(k)		*name(parameters)*
resources		$\{r_k^1, r_k^2, \ldots, r_k^r\}$
if		Φ^c
and if not		Φ_k^e
and if		Φ_k^p
and if not		Φ_k^f
then		
	do	$(a_k^1, a_k^2, \ldots, a_k^a)$
	retract	$\{d_k^1, d_k^2, \ldots, d_k^d\}$
	assert	$\{h_k^1, h_k^2, \ldots, h_k^h\}$
	output	$(m_k^1, m_k^2, \ldots, m_k^m)$
next		(n_k)
else		(e_k).

In the above specification k is the rule number (unique for any rule). The resources necessary to apply rule k are denoted with $\{r_k^1, r_k^2, \ldots, r_k^r\}$, Φ^c is the formula defining the context, Φ_k^e is a formula describing the initial excluding conditions for fast pruning of candidate transformation rules, Φ_k^p is a formula describing the preconditions necessary for application of the rule (typically, it is a DNF or simple conjunctive formula for the situation consisting of all the states in which rule k can be applied), Φ_k^f is another auxiliary formula describing excluding conditions for refined pruning of certain transformation rules, $(a_k^1, a_k^2, \ldots, a_k^a)$ is the sequence of actions to be executed, $\{d_k^1, d_k^2, \ldots, d_k^d\}$ is the set of delete conditions (in the form of literals) which are no longer true after application of the rule, $\{h_k^1, h_k^2, \ldots, h_k^h\}$ is a set (a simple formula) containing all the literals which become true as a result of application of the rule and are to be asserted to the knowledge base, $(m_k^1, m_k^2, \ldots, m_k^m)$ are the messages to be sent to the console, and finally (n_k) is the number of the next candidate rule which should be checked first for possible application after rule k is fired; if it is impossible to fire the rule, rule number e_k should be checked next.

The generic algorithm for rule application is as follows:

1. Check if all the resources $\{r_k^1, r_k^2, \ldots, r_k^r\}$ (if specified) are available. If yes — proceed. If no — exit, the rule cannot be applied. Go to rule e_k if specified.
2. Check if Φ^c is satisfied in the state of the knowledge base defined by ϕ, i.e. if there is $\phi \models \Phi^c$. If yes — proceed. If no — exit, the rule cannot be applied. Go to another context.
3. Check if Φ_k^e is not satisfied i.e. $\phi \not\models \Phi_k^e$ (if Φ_k^e is specified). If it is unsatisfied — proceed. In the other case — exit, the rule cannot be applied. Go to rule e_k if specified.
4. Check if Φ_k^p is satisfied, i.e. if $\phi \models \Phi_k^p$. If yes — proceed. If not — exit, the rule cannot be applied. Go to rule e_k if specified.
5. Let σ be the substitution obtained during the check in the preceding step, i.e. one specifying all the necessary replacements of variables by terms in Φ_k^p. Check if $\phi \not\models \Phi_k^f\sigma$. If $\Phi_k^f\sigma$ is not satisfied — proceed. In the other case — exit, the rule cannot be applied. Go to rule e_k if specified.
6. Apply all the actions defined by $(a_k^1, a_k^2, \ldots, a_k^a)$ after substituting for any variables the terms indicated by σ.
7. Delete from the current state formula all the facts matched by any of the facts of $\{d_k^1, d_k^2, \ldots, d_k^d\}\sigma$.
8. Assert to the state formula all the facts included in $\{h_k^1, h_k^2, \ldots, h_k^h\}\sigma$.
9. Send to the output all the messages given by $(m_k^1, m_k^2, \ldots, m_k^m)$, after substituting for the variables occurring in the messages the terms indicated by σ.
10. Return control to the upper level mechanism, passing the next rule number (n_k) of a rule to be examined first (if specified).

The above algorithm outlines the generic one-step inference process, i.e. application of a single extended rule to the current knowledge-base. During the execution a general theorem proving mechanism may be necessary. In case of complex rules one may apply the resolution theorem proving or theorem proving based on backward dual resolution. However, in typical cases theorem proving is limited to formulae matching.

In most of practical applications, the Φ rules (defining context, excluding conditions and preconditions) are limited to be simple conjunctive formulae. In such cases the check for logical consequence can be performed by matching the components (literals) of the formula to be proved against the components of the knowledge base. Such a procedure consists of searching for an appropriate substitution for direct proving of preconditions by appropriate instantiation of rule parameters.

Note that if the *next* and the *else* part are specified, then in fact there is no problem of conflict resolution provided that the initial rule to start with is defined. In fact, it is not necessary to always define the control links in an explicit way. If, for example the rules are linearly or hierarchically ordered, then the lack of specification of the *next* or the *else* rule may be interpreted by default as the request to go to the next rule within the order. Hence, the proposed rule format together with the interpretation algorithm may solve the conflict resolution problem in an operational way.

10.3 Inference Control at the Rules Level: Advanced Problem

In this section some selected strategies of conflict resolution are outlined. A good overview of some most popular approaches for control systems is provided in [130]; the following strategies are listed:

1) rule ordering (a rule appearing earliest has the highest priority),
2) data ordering (a rule with the highest priority assigned to its conditions has the highest priority),
3) size ordering (a rule with longest list constraining conditions has the highest priority),
4) specificity ordering (arrange rules whose conditions are super-set of another rule,
5) context limiting (activating or deactivating groups of rules at any time to reduce the occurrence of conflict).

The two first approaches belong to direct, static ordering strategies. The third and the fourth ones constitute some specific cases of ordering possibilities with mostly syntax-based priority evaluation mechanism. Only the fifth one refers to semantics of the process but it is used to *reduce* the number of rules in the conflict set, and not to select a unique rule.

In OPS5 rule-based programming system [12] two more complex strategies, LEX and MEA, can be used. Roughly speaking, these strategies are implemented in four subsequent steps: *refraction* (deleting all instantiations fired previously), partial ordering based on *recency* (time tags corresponding to the working memory elements used are considered), partial ordering with respect to *specificity* (rules with greatest number of tests are considered first), and, finally, arbitrary selection. The MEA strategy considers the recency of the first condition of rules most important. Note that strategies as the one used for rule-based programming may be unsatisfactory or even inadmissible for control of dynamic systems. For example, the refraction strategy may remove some instantiation previously executed, while repetition of exactly the same rule may be necessary to achieve certain goal via several similar moves.

Other strategies and problems of conflict resolution with respect to expert systems are also discussed in [44, 46, 51]. Problems of real-time rule-based control issues referring to rule selection and control strategies are also discussed in [48, 142]. In [53] a selection of reasoning control mechanisms with short discussion are given.

10.3.1 A Simple Linear Strategy

Below, a simple linear strategy for conflict resolution is outlined.

Let us assume that in a knowledge based system (or its component designated to work in a given context) there are as many as n rules specified. The reasoning control mechanism repeatedly performs the following actions:

- matches all the rules (in a certain established order) against the current state formula;
- selects a subset of all the rules which satisfy the applicability conditions, i.e. generates the so-called *conflict set* (the rules are selected with regard to the points 1–4 of the algorithm presented in Sect. 10.2);
- resolves the conflict by selecting a single rule from the conflict set;
- applies the selected rule with regard to the points 5–8 of the algorithm presented in Sect. 10.2;
- creates a hierarchy of rules for the next iteration with regard to information given in the **next** part of the executed rule in the algorithm presented in Sect. 10.2;
- goes to another iteration.

The basic assumption underlying the implementation of the above reasoning control mechanism is that at any stage only a single rule can be applied. This is a natural assumption, since any rule changes the state of the system in an independent way. However, in certain practical implementations, if the rules do not affect directly the process, it is possible to select more than one rule from the conflict set, 'combine' together the results of their applications, and finally execute the resulting (from the combined rules) action[1]. However, for the sake of simplicity we shall only consider the case of a single rule selection and application at a time.

Note that there are two basic possibilities. First, it may be the case that all the rules are mutually exclusive, i.e. the satisfaction of applicability conditions

[1] This in fact is the case of fuzzy controllers.

for one of the rules automatically excludes all the other rules. In such a case we shall say that the system is *deterministic*. We shall discuss such systems more precisely in one of the next chapters. Here note only that in such a case there is no need to apply any conflict resolution mechanism.

In the more complicated case, the conflict set may contain two or more potentially applicable rules. Now, the task of the reasoning control mechanism is to select and apply one of them. The above problem has no unique solution.

In practice, various approaches using different auxiliary information are applied [130]. The most popular ones include the following:

- ordered execution of the rules from the knowledge base (only those of them which have the conditional part satisfied); typically a linear order of execution is followed; this include the case of the additional information placed in the **next** part of the rules — the specified rules are examined first, and the first one selected is applied; this is aimed at speeding up the execution of some typical routines;
- various priority evaluation mechanisms can be applied so as to establish the order of rules in the conflict set;
- a meta-knowledge system for selection of the rule to be applied can be used; such a system can use typical expert systems approach and take into account the current goal of reasoning, certain environmental conditions, some statistics concerning the past system behavior, etc.;
- certain facts can be used as semaphores for blocking and enabling the inference control mechanism to limit the subset of rules which can be executed next.

In fact, the choice of detailed conflict resolution mechanism must depend on the domain-specific knowledge — no general solution seems to be optimal.

11

Logic Programming and Prolog

Logic Programming constitutes perhaps one of the most brilliant ideas among the concepts for programming languages, knowledge representation and inference. The idea is originally due to Robert Kowalski from the Kings College in London [47]. It consists of direct application of a subset of First-Order Logic for declarative encoding of knowledge and application of a specific strategy of resolution theorem proving for inference. All together — declarative specification of user knowledge joined with automated reasoning paradigm creates an interesting and powerful programming language.

PROLOG itself is perhaps the most beautiful, simple, yet powerful programming language ever created by man. In its pure form there are no 'instructions' — contrary to classical programming languages, there are no explicit constructions such as **begin**...**end**, **goto**, **if**...**then**...**else**, **case**...**of**...**end**, **while**...**do**, **for**...**do**..., etc. Instead, PROLOG uses practically unchanged syntax of First-Order Predicate Calculus, which allows for specification of rules in a similar to implication form.

And this is perhaps the most significant characteristic of PROLOG — the code in this language is practically composed from one type of rules, which from logical point of view are Horn clauses.

The distinctive features of PROLOG make it a noble, distinguished tool, taking a very specific position among — or perhaps above — other, even most advanced programming languages.

Among these features, the following ones seem to be of primary importance:

- *declarative language* — PROLOG is a *declarative* programming language versus most of the other languages being *procedural* ones;
- *relational language* — it is based on the concept of *relation* rather than function; certainly, it is not a functional language, and hence, in its pure form 'procedures' have no clearly defined input and output;

A. Ligęza: *Logical Foundations for Rule-Based Systems*, Studies in Computational Intelligence (SCI) **11**, 173–188 (2006)
www.springerlink.com

- *multi-level language* — PROLOG programming can be done at different levels — knowledge can be encoded directly as logical clauses or by *meta-programming* one can specify both knowledge and inference rules;
- *internal mechanisms* — PROLOG possesses three strong internal mechanisms build into the inference engine; these are *unification*, *resolution*, and *SLD inference strategy*;
- *no instructions* — in the pure PROLOG practically no instructions in the classical sense are necessary.

Let us explain the intriguing features in detail.

First, PROLOG is a declarative language, at least in its pure logic-programming form. This means that one has to specify his knowledge concerning specific domain, and then questions can be asked; the answers to them are deduced internally, and the user need not bother how it is done. By no means he has to program procedures for calculating the answers. This is a very important idea; in practical applications, however, a knowledge about controlling the inference sometimes has to be also specified as a part of the program code.

Second, PROLOG is based on the concepts of *relation* and *logical formula* rather than the one of a *function*. This means that specification of knowledge takes the 'relational' form. Predicates specifying relations can have numerous arguments, but these arguments are not divided into input ones and output ones; in principle, any of them can play both the roles, depending on the current task to be solved.

Third, programming in PROLOG can be performed at direct level, when knowledge is encoded directly in the form of logical clauses to be interpreted according to the built-in strategy or, using the idea of term, arbitrarily complex knowledge structures can be encoded and using rule-based programming any specific inference mechanism can be specified. Moreover, PROLOG can manipulate its own code and in this way one can build self-modifying programs.

A unique, distinctive feature of PROLOG is constituted by incorporation of three very strong mechanisms in it. First, *unification* allows for term matching by finding unifying substitutions. Arbitrarily complex structures can be quickly compared and this feature allows for pattern matching when no programming effort is required. Second, from facts and clauses new facts and clauses can be deduced via resolution method. This means that new knowledge, which is logical consequence of the one already specified can be produced in an automatic way. Third, a depth-first search procedure with backtracking mechanisms is incorporated in any implementation of PROLOG which means that search for solution can be accomplished by simple question asking. These three mechanisms makes PROLOG programming extremely efficient — in order to replace a few lines of its code one must use sometimes hundreds of lines of other, even high-level languages.

Finally, in pure logic programming, and in fact — in core of the applications built in PROLOG — no 'classical instructions' are necessary. This follows from the fact that PROLOG is a declarative language and that the necessary interpretation skills are already built-in. In fact, although implementations of PROLOG can provide even around thousand specific predefined predicates, one can start using this nice language without even looking at them.

There exist numerous books on PROLOG. A concise but in-depth introduction to logical foundations of the language and its mechanism and perhaps the best theoretical book is the one by Nilsson and Małuszyński [105]. A very interesting, direct introduction to PROLOG applications in AI is the book by Bratko [11]. There are many other interesting positions, such as [10, 19, 24, 82, 84, 121]. In the next sections we present a short introduction and show some solutions for development of rule-based systems in PROLOG.

11.1 Introductory Example

In order to quickly provide some intuitions on how PROLOG code looks like and what PROLOG programming consists of let us have a look at the following simple code.

```
male(adam).
male(bill).
male(chris).
male(danny).

female(ann).
female(mary).

father(adam,bill).
father(adam,chris).
father(adam,danny).
father(adam,ann).
father(adam,mary).

brother(X,B) :-
   father(F,X),
   father(F,B),
   male(B),
   B \== X.
```

A PROLOG program consists of *facts* and *rules*. In the presented program there are 11 facts and one rule. Specification of any fact is composed of the name of the relation followed by its arguments (placed in parentheses, separated by comma). So for example `male(adam)` means 'adam is a male', `female(ann)` means 'ann is female', and `father(adam,bill)` means 'adam is the father of bill'. A predicate symbol p having n arguments is denoted as p/n.

In general, there are the following rules concerning facts:

- fact has always a name of the relation,
- it can have arbitrary number of arguments (including zero),

- it is followed by full stop,
- facts having the same name (predicate symbol) are put together in the code,
- facts having the same name but different number of arguments are simply different facts belonging to different groups.

Note also, that according to the convention names (constants) begin with lower case letter, so we write 'adam' instead of 'Adam'. Strings starting with upper case letters are variables.

In the program above there is also one rule. Using basic facts one can construct inference rules which in PROLOG are called *clauses.* The only clause in the code above states that B is the brother of X *if* F is the father of X, the same F is the father of B, B is a male and B is different than X.

The formula `brother(X,B)` is called the HEAD of the clause, while the other atoms forming its definition are called the *body* of the clause. Instead of '`:-`' one can also write (and read) *if*, and instead of the commas separating the components of the body one can write (and read) *and.* Any clause is ended with a full stop.

In this way one can specify practically arbitrary complex knowledge concerning any domain. In order to run PROLOG programs one has to ask questions and read the answers generated by the interpreter. For example, in the case of the above program one can ask the following questions:

`?- male(adam).`	—	The answer is '`yes`'.
`?- male(X).`	—	The answers are: '`X=adam`', '`X=bill`', '`X=chris`', and finally '`X=danny`'.
`?- father(adam,bill).`	—	The answer is '`yes`'.
`?- father(bill,adam).`	—	The answer is '`no`'.
`?- father(X,Y).`	—	There will be five different answers where always '`X=adam`' but `Y` will take values '`bill`', '`chris`', '`danny`', '`ann`', and '`mary`'.

In the above, '`?-`' is the classical prompt of PROLOG interpreter. In general, if a ground fact can be found in the program, the answer is '`yes`'; in the other case it is '`no`'. If the facts contain variables, the interpreter tries to find any possible substitution for these variables so that a fact can be made true with respect to the provided knowledge. To get all the answers one normally has to prompt the interpreter by pressing the '`;`' character (in logic programming it plays the role of logical OR connective).

The most interesting, however, is the case of facts (with or without variables) which cannot be found directly in the program. In this case the interpreter tries to *deduce* them with the use of the clauses. Without going into details, the procedure consists of looking for all the clauses allowing to deduce a given goal fact. If the fact can be unified with the head of a clause, its proof is replaced by a proof of the body of this clause.

In our example, to find all the brothers of bill we have to ask a question of the form: `brother(bill,B)`. The interpreter unifies this goal with the head of the clause; since it works, now the goal is to prove that bill and B have the same father F (it works by proving `father(F,bill)` and `father(F,B)` for F being 'adam', while B takes five different values. From these values only three satisfy the condition `male(B)`, and only two of them are different from 'bill'. So the final answers found are '`B=chris`' and '`B=danny`'.

The above example, although a very simple one, constitutes in fact a *deductive database*. Facts are analogous to records in relational database. Groups of facts with the same predicative symbol (and number of arguments) correspond to tables of relational databases. Asking simple questions about such facts is very much like performing the search on a relational database. Asking more general questions, where to generate the answer deduction is required, has no direct interpretation in classical databases. In fact, it is a powerful operation of generating new knowledge from facts with the use of inference rules.

11.2 Prolog Syntax

The basic syntax of PROLOG is very much the same as the one of First Order Predicate Calculus, but restricted to the Horn clauses only. All variables are universally quantified (by default; no explicit quantifiers are used). Variable names start with an upper case letter or an underscore. There is also the so called *universal variable*, symbolized with '_' (underscore), unifiable with any term.

Objects are represented with terms, exactly as in the case of First Order Logic. Terms were defined and discussed in Sect. 2.2. Let us recall shortly that a term is any constant, any variable, and if f is a functional symbol of arity n, then $f(t_1, t_2, \ldots, t_n)$ is also a term, provided that $t_1, t_2, \ldots, t_n$ are terms.

The basic concept for knowledge representation in PROLOG are facts. From logical point of view facts are atomic formulae. Any fact in PROLOG is followed by full stop.

Atomic formulae were discussed and defined in Sect. 2.3. Let us recall that an atomic formula is always of the form $p(t_1, t_2, \ldots, t_n)$, where p is predicate symbol of arity n and $t_1, t_2, \ldots, t_n$ are terms. To denote the fact that the number of arguments of p is n we use the notation p/n.

The only more complex logical formulae in PROLOG are Horn clauses. They were defined in Sect. 2.4. Let us recall that a Horn clause is a disjunction of literals with at most one positive literal. Let $p_1, p_2, \ldots, p_k$ be positive literals. A *Horn clause* is any clause of the form

$$\neg p_1 \vee \neg p_2 \vee \ldots \neg p_k \vee q,$$

where q is a literal (either positive or negative one).

In other words, a Horn clause is any clause with at most one positive literal. By the De Morgan's laws and the definition of implication, a Horn clause can be represented as an implication of the form

$$p_1 \wedge p_2 \wedge \ldots \wedge p_k \Rightarrow q.$$

Thus, a Horn clause can be regarded as an *inference rule*, having preconditions defined by $p_1 \wedge p_2 \wedge \ldots \wedge p_k$ and conclusion q.

Assume q is a positive literal. Using the conventions of logic programming the above clause can be presented as:

$$q \Leftarrow p_1 \wedge p_2 \wedge \ldots \wedge p_k.$$

In practice, in PROLOG notation it is written as:

$$q \quad \textbf{if} \quad p_1 \quad \textbf{and} \quad p_2 \quad \textbf{and} \quad \ldots \quad \textbf{and} \quad p_k$$

or for the sake of conciseness as:

$$q \;\texttt{:-}\; p_1, p_2, \ldots, p_k.$$

Clauses with the same head predicate are grouped together and they constitute a subprogram or a block of clauses. All variables are local in a clause.

11.3 Unification in Prolog

Unification is accomplished by finding replacement of variables with terms so that two or more terms (atomic formulae) become identical. The specification what term should be substituted for which variable is called a *substitution.* Recall that any substitution σ can be presented as

$$\sigma = \{X_1/t_1, X_2/t_2, \ldots, X_n/t_n\},$$

where t_i is a term to be substituted for variable X_i , $i = 1, 2, \ldots, n$. If Φ is a formula (or term) and σ is a substitution, then $\Phi\sigma$ is the formula (or term) resulting from replacing the variables of Φ with the appropriate terms of σ.

The details on substitution and unification were covered in Sect. 4.1. A universal algorithm for unification of terms (or atomic formulae) was presented there in Subsect 4.1.3.

The unification algorithm in PROLOG is practically identical, apart from the so-called *occur-check.* In step 4 of the algorithm there is a check if the variable to be substituted for does not occur in the term being substituted (if so, we obtain in fact a cyclic expression, and both its usefulness and the dealing with it become somewhat problematic). As this check is time-consuming (due to its computational complexity) most of the practical implementations of PROLOG simply omit this step. It is thus the user's task to write the program in such a way, that occur-check is no longer necessary. Further, if one wants, one may require the use of full algorithm with occur-check (if provided by the implementation).

Practically, unification algorithms are usually very fast and they can operate on arbitrarily complex structures, provided that the scheme of a term is kept. The user does not perceive the work of unification mechanism unless a special `trace` mode is switched on.

11.4 Resolution in Prolog

Resolution is the basic step of inference during execution of a PROLOG program.

The logical foundations of resolution based inference and resolution theorem proving were presented in Sect. 4.3. In PROLOG, resolution is applied at any step of inference — from the clause defining the current goal and a selected input clause coming from the program code, a new clause is deduced. Below a scheme of this basic step is recapitulated.

Let P denote the PROLOG program — it can be considered to be a conjunction of all its facts and clauses, so it constitutes a complex logical formula. The question stated to the interpreter and constituting the current goal is a formula to be proved from P; let us denote the current goal as G. In fact, the logical task is to show that

$$P \models G, \tag{11.1}$$

i.e. that the goal G is a logical consequence of P.

In resolution theorem proving all the proofs are based on refutation. Hence instead of (11.1) the task is stated so as to show that

$$P \wedge \neg G \tag{11.2}$$

is unsatisfiable. In fact, the goal, from logical point of view is considered as negated initial formula, added to the initial knowledge base as a special clause, and it is attempted to prove that such a combination is unsatisfiable. This is carried out by generating an empty formula as a consequence of (11.2).

Now, let $q_1, q_2, \ldots, q_n$ be the set (logically: conjunction) of goals to be proved at a certain step of reasoning. In logical convention, since all the proofs with resolution are based on refutation, one has to prove that

$$\neg(q_1 \wedge q_2 \wedge \ldots \wedge q_n) \wedge P$$

is unsatisfiable.

After applying the De Morgan's law, the current goal can be represented as a Horn clause of the form

$$\neg q_1 \vee \neg q_2 \vee \ldots \vee \neg q_n. \tag{11.3}$$

Let

$$h \texttt{ :- } p_1, p_2, \ldots, p_m.$$

be the selected program clause. Obviously, the aim is to apply resolution so that a new goal is generated.

In logical notation the selected clause takes the form of a Horn clause, i.e.

$$h \vee \neg p_1 \vee \neg p_2 \vee \ldots \vee \neg p_m. \tag{11.4}$$

PROLOG interpreters work in such a way that it is always the first (from the left) goal to be unified with the head clause. Hence, if σ is the most general unifier of q_1 and h then a new clause can be generated be resolving clauses (11.3) and (11.4). Formally, the resolution rule applied in PROLOG takes the following form

$$\frac{\neg q_1 \vee \neg q_2 \vee \ldots \vee \neg q_n\,,\; h \vee \neg p_1 \vee \neg p_2 \vee \ldots \vee \neg p_m}{(\neg p_1 \vee \neg p_2 \vee \ldots \vee \neg p_m \vee \neg q_2 \vee \ldots \vee \neg q_n)\sigma}. \tag{11.5}$$

In practice, a new set of goals

$$p_1\sigma, p_2\sigma, \ldots, p_m\sigma, q_2\sigma, \ldots, q_n\sigma$$

is produced.

Note that the goal of such inference is to eventually end up with an empty clause. The length of the goal sequence can be reduced only if resolving with facts (a fact can be considered as a clause with empty body). Resolving with a fact always reduces the length of the goal by one.

11.5 Prolog Inference Strategy

In fact the basic inference strategy implemented in standard PROLOG is quite simple. It is the so-called SLD-Strategy, which means that PROLOG uses Linear refutation procedure for Definite clauses with Selection function.

Linear strategy means that the new resolvent can be produced only from the one produced a step earlier; in fact the sequence of resolvents generated during an attempt to produce the empty formula forms a line. At any step, the current resolvent defining the goal is combined with one of the input program clauses (a clause or a fact), so it is also an *input* strategy.

The selection function is a function choosing a subgoal from the current resolvent to be used next. In standard implementations it is simple — always the leftmost goal is processed first.

When looking for facts and clauses for one unifiable with the selected subgoal the natural top-down ordering is assumed. Hence, although from logical point of view a program can be considered as a conjunction of clauses (the order is unimportant), the order of clauses in a program may have significant influence on the execution of practical programs.

PROLOG interpreters explore all the solutions, so whenever a solution is found (an empty clause is generated) the interpreter backtracks and explores other possibilities. They are provided by different possibilities of selecting the clause for resolution.

To summarize, the outline of the concept of PROLOG interpreter is as follows.

1. Let R be the current resolvent defining the goal; for the beginning let $R = G$, where G is the user-defined goal. Further, initialize stack memory

and empty it (the stack will be used to remember alternative resolving possibilities).

2. If the current resolvent R is empty and the stack is empty then exit with the output 'True' with the values of parameters of G; the program ends with success since all the subgoals have been solved. No other inference possibilities are available.
3. If the current resolvent R is not empty, $R = q_1, q_2, \ldots, q_n$ then:
   - take and remove from R the first subgoal, i.e. q_1;
   - find the first clause h :- $p_1, p_2, \ldots, p_m$ such that q_1 is unifiable with h; let the mgu be σ;
   - put on the stack any other clauses having head unifiable with q_1 together with the current goal definition;
   - generate the new resolvent R' according to resolution rule (11.5);
   - put $R = R'$ and go to 2.
4. If the current resolvent is empty but the stack contains other reasoning possibilities then output 'True' with the values of parameters of G. Backtrack by removing the first clause and the corresponding goal from the stack; put the goal for R and go to 2.
5. If the current resolvent is not empty, but it is impossible to unify the first subgoal with any fact or clause head then:
   - if the stack is not empty, backtrack and proceed as in 4;
   - if the stack is empty output 'False' (if no solution was found earlier) and exit.

In fact, PROLOG interpreter performs a search for solutions. It employs the depth-first search strategy with backtracking any time a solution is found or no further resolvent can be found. The stack memory helps to organize the search; on the top of the stack a new alternative search possibility is always ready to explore (if only the stack is not empty).

11.6 Inference Control and Negation in Prolog

The search and inference strategy implemented in PROLOG do a big work for programmers; not only unification algorithm allows to compare structural objects, but the resolution allows for exploring very long inference paths in an automatic way. Further, the incorporated search strategy allows for exploration of all the inference possibilities and thus all the solutions are found.

In practice, it may be useful to control the inference in a limited way. There are two intrinsic standard predicates built in any PROLOG implementation:

1) the *cut* predicate,
2) the *fail* predicate.

They are described below together with the *not* predicate which allows to implement negation through the so-called *Closed-World Assumption* (CWA) [39].

11.6.1 The *cut* Predicate

The *cut* predicate is a standard predicate which helps avoiding exploration of further inference possibilities, provided that the one currently explored satisfies certain predicates. The cut predicate allows for pruning branches in the search tree generated by the depth-first search algorithm. In practice, the *cut* predicate is symbolized with the exclamation mark '!'.

Consider a clause of the form as below

$$h \text{ :- } p_1, p_2, \ldots, p_i, p_{i+1}, \ldots, p_m.$$

During execution of PROLOG program it is possible to backtrack at any atomic formula p_i in the body of the clause, as well as for different variants of h. Consider a clause

$$h \text{ :- } p_1, p_2, \ldots, p_i, !, p_{i+1}, \ldots, p_m.$$

with the *cut* predicate placed between atoms p_i and p_{i+1}. The *cut* divides the clause into two parts, the one to the left of cut and the one to the right of it. The operation of cut is simple: if the left-located atoms are proved, then by 'passing through' the cut all other possibilities for them are removed from the stack memory. This means that during backtracking, there will be no exploration of other variants for h, p_1, $p_2, \ldots, p_i$. On the other hand, backtracking is still possible for goals defined with $p_{i+1}, \ldots$, up to p_m. Hence *cut* predicate can be used to restrict the search by avoiding exploration of all the inference possibilities.

11.6.2 The *fail* Predicate

The *fail* predicate has a very simple role — it stops inference and forces backtracking. This is so because *fail* is a predicate such that it is impossible to unify it with any other predicate. So, for example, to ensure exploration of all the possible executions of a clause we often place *fail* at the end of this clause.

Practical applications of *fail* include implementation of loops in PROLOG. A typical scheme of such a loop is as follows.

```
loop :-
    action,
    fail.
loop.
```

In the construction above, after calling `loop` the operation defined by `action` is executed. If there are any other possibilities of executing it, then after `fail` they will be explored. In fact, this may be an infinite loop if there are infinitely many possibilities of executing `action`. In case there are no other such possibilities, the second clause for `loop` is executed; in practice it does nothing, but it is necessary here to 'close' the loop, since without it the execution of the loop would end up with a failure.

11.6.3 The *not* Predicate

Predicate *not* is in fact a meta-predicate — its argument is a predicate p, and $not(p)$ succeeds if an attempt to prove p fails. Hence, *not* is a predicate which implements *negation as a failure.*

The concept of negation in PROLOG is based on the so-called *Closed-World Assumption.* The idea is as follows: positive knowledge about the world of interest is stated explicitly. It takes the form of facts and rules (clauses in PROLOG). It is assumed that *all* the positive knowledge is available or can be deduced; in other words, that the world is closed. In this case, if a certain fact does not follow from the current state, it is assumed to be false. It does not mean it is really false; the definition is *operational* — it provides a way to decide whether something is false in the case the negation of it is not stated in an explicit way.

Note that the definition of *not* can be expressed in PROLOG-style code as follows.

```
not(P):- P,!,fail.
not(_).
```

In the above code, if P succeeds, the execution of *not* fails; since in the first clause *cut* is used, the second clause is not executed. On the other hand, if an attempt to satisfy P fails, its negation can be assumed true thanks to the second clause.

11.7 Dynamic Global Memory in Prolog

Programs in PROLOG can easily access global memory. There are two basic operations on it — one can *assert* new facts and the other can *retract* them. The standard predicates *assert/1* and *retract/1* are in fact meta-predicates; their arguments are facts.

The use of the above predicates is simple. In order to assert new fact p to the global memory one uses the construction

$$assert(p).$$

In order to retract p from the global memory one uses the construction

$$retract(p).$$

Predicates *assert* and *retract* can be used in any place and they constitute a part of the goal or body of certain clauses. At any stage of program execution all the clauses have access to the knowledge contained in the global memory; a useful mechanism for communication among clauses without parameter passing is available in this way. The *retract* and *assert* operations produce results which are not removed during backtracking.

In order to clear the memory from all occurrences of predicate p containing some parameters one has to repeat the retract operation until there is nothing to be retracted. The predicate which performs clearing of the memory is called *retractall/1*; its definition is as follows:

```
retractall(P):-
   retract(P),
   fail.
retractall(_).
```

One can notice that the defined as above *retractall* is a typical PROLOG loop. Most of the implementations of PROLOG provide this predicate as a standard, built-in procedure.

11.8 Lists in Prolog

Lists are one of the most important structures in symbolic languages. In most of the implementations of PROLOG lists are standard structures and there are numerous operations on them provided as built-in procedures. Lists can be used to represent sets, sequences, and more complex structures, such as trees, records, etc.

A list in PROLOG is a structure of the form

$$[t_1, t_2, \ldots, t_n].$$

The order of elements of a list is important; the direct access is only to the first element called the *head*, while the rest forms the list called the *tail*.

Lists in fact are also terms, and a list as above is equivalent to a term defined as follows:

$$l(t_1, l(t_2, \ldots l(t_n, nil) \ldots)),$$

where l is the list constructor symbol and nil is symbolic denotation of an empty list.

In practical programming it is more convenient to use the bracket notation. In order to distinguish the head and the tail of a list the following notation is used

$$[H|T].$$

If $L = [a, b, c, d]$ then after unifying L with $[H|T]$ we obtain $H = a$ (a single element), and $T = [b, c, d]$ (a list).

A list can have as many elements as necessary. An empty list is denoted as $[\,]$. A list can have arguments of complex structures, i.e. terms, lists, etc.

Considering a list as a set one can define the following two important basic operations.

```
member(Element,[Element|_]):- !.
member(Element,[_|Tail]):- member(Element,Tail).

select(Element,[Element|Tail],Tail).
select(Element,[Head|Tail],[Head|TaiE]):- select(Element,Tail,TaiE).
```

The operation of the above predicates is simple and similar; the basic idea is common. The *member/2* predicate checks if the first argument belongs to the set represented by the second argument. This may be so if it is the first element of the list or if it is a member of the tail of the list; the definition is recursive. The *select/3* predicate operates in a similar way, however, its primary use consists in selecting an element of a set and returning the set without the currently selected element. In fact, the *select/3* predicate can be considered as indeterministic choice — after backtracking it returns different element of the set each time.

11.9 Rule Interpreters in Prolog

PROLOG is one of the most elegant programming languages, but it is also one of the most efficient ones. During designing the code, the programmer can concentrate on the logical aspects of the task and not on the technical details of the code.

Below, we shall show two very simple examples of rule interpreters written in PROLOG. The main goal is to show the way one can use PROLOG as a meta-interpreter.

Consider the following example table specifying the functional behavior of full adder. The signals p and q are the inputs, c is the carry from previous unit, s is the bit of the output and c' is the output carry signal (Table 11.1).

Table 11.1. A table defining the full adder

No.	p	q	c	s	c'
1	0	0	0	0	0
2	0	0	1	1	0
3	0	1	0	1	0
4	1	0	0	1	0
5	0	1	1	0	1
6	1	0	1	0	1
7	1	1	0	0	1
8	1	1	1	1	1

Assume we would like to specify the behavior of the system with simple rules of the form:

```
rule(<rule_number>,
     <positive_preconditions>,
     <negative_preconditions>,
     <retract_facts>,
     <assert_facts>,
     <output>).
```

The above behavior can be specified with the following set of rules:

```
rule(1,[],[p,q,c],[negative(p),negative(q)],[],[negative(s)]).
rule(2,[c],[p,q],[negative(p),negative(q),positive(c)],
     [negative(c)],[positive(s)]).
rule(3,[q],[p,c],[negative(p),positive(q)],[],[positive(s)]).
rule(4,[p],[c,q],[negative(q),positive(p)],[],[positive(s)]).
rule(5,[q,c],[p],[negative(p),positive(q)],[],[negative(s)]).
rule(6,[p,c],[q],[positive(p),negative(q)],[],[negative(s)]).
rule(7,[p,q],[c],[positive(p),positive(q),negative(c)],
     [positive(c)],[negative(s)]).
rule(8,[p,q,c],[],[positive(p),positive(q)],[],[positive(s)]).
```

The system is initiated by providing the initial values of the lowest bits of the numbers to be added by asserting them into the dynamic memory as for example:

```
assert(positive(p)).
assert(negative(q)).
assert(negative(c)).
assert(position(1)).
```

The system is to react — for the given initial state only rule number 4 can be executed. As the results of application of the rule the output for that bit is produced as *positive*(s) and the carry bit remains unchanged. Then the next input bits are put into the memory and the cycle is repeated.

The code of the rule interpreter is as follows:

```
run(N):-
   retract(position(K)),
   K =< N,
   rule(_,Preconditions,NegativePreconditions,Retracts,Asserts,
                                                         Outputs),
   satisfied(Preconditions),
   unsatisfied(NegativePreconditions),
   remove(Retracts),
   add(Asserts),
   output(Outputs),
   !,
   K1 is K + 1,
```

```
   assert(position(K1)),
   run(N).

run(_):- clean.
clean:-
    retractall(positive(_)),
    retractall(negative(_)),
    retractall(position(_)).
```

The interpreter starts with checking the bit position; after any cycle the position is increased by 1 until it becomes bigger than the length of the added digits; at the end of the run the memory is cleaned from positive and negative facts. The basic cycle is as follows: a rule is read, it is checked if its positive preconditions are satisfied and simultaneously the negative ones must be dissatisfied, and if so the memory is modified by retracting the old signals p and q, and changing c if necessary (in fact it is performed by rule 2 and 7 only). The current bit with result is send to the output (here written). After execution of the rule the stack is cleaned with the *cut/1* predicate, the position number is increased by 1 and the cycle is repeated.

Note that the inference control strategy is the *closed-loop linear hierarchical control strategy*, as described in Sect. 10.1.1

Note that, if properly initiated (all the signals and the initial bit position must be defined) the system always finds a rule to be fired (it is *complete*) and there is exactly one such rule (it is *deterministic*). It is complete, since there are exactly 3 input binary signals and 8 (2^3) rules covering all possible input states. Further, the preconditions of the rules are mutually exclusive, and so the system is deterministic. The presented interpreter performs finitely many times a loop of the type 'end-calls-beginning' with stack memory cleaning.

The auxiliary predicates are all listed below.

```
satisfied([]):- !.
satisfied([H|T]):-
   positive(H),
  satisfied(T).
unsatisfied([]):- !.
unsatisfied([H|T]):-
   negative(H),
   unsatisfied(T).

remove([]):- !.
remove([H|T]):-
   retractall(H),
   remove(T).

add([]):- !.
add([H|T]):-
   assert(H),
   add(T).
```

```
output([]):- !.
output([H|T]):-
   write(H),nl,
   output(T).
```

All of the predicates constitute typical recursive definitions of operations on lists. Note that they are of generic nature; by modifying the definitions of *positive/1* and *negative/1* one can easily redefine them so that they operate on attributive or even first order logic knowledge base.

Note that for various applications various inference control mechanisms are adequate. It is easy to modify the scheme, so as to obtain various control effects. For example, if one wants to execute all the rules with satisfied preconditions, i.e the *closed-loop linear strategy* (see Sect. 10.1.1), one can modify the loop so that the 'repeat-fail' scheme is used.

```
run(N):-
   rule(_,Preconditions,NegativePreconditions,Retracts,Asserts,
                                                          Outputs),
   retract(position(K)),
   K =< N,
   satisfied(Preconditions),
   unsatisfied(NegativePreconditions),
   remove(Retracts),
   add(Asserts),
   output(Outputs),
   K1 is K + 1,
   assert(position(K1)),
   fail.
run(_):- clean.
clean:-
    retractall(positive(_)),
    retractall(negative(_)),
    retractall(position(_)).
```

PROLOG will be applied to show how rules and rule interpreters can be implemented. More complex interpreter will be shown in one of the further sections concerning design principles. It will also be applied to illustrate how to construct modules for verification of rules.

Part III

Verification of Rule-Based Systems

12

Principles of Verification of Rule-Based Systems

Designing and implementation of rule-based systems is frequently considered as a specific activity close to classical computer programming; however, there are specific differences which make this activity a somewhat special domain which requires separated approach and special attention must be paid to selected issues. The most important differences with respect to classical programming languages include the following features:

- design and development of a rule-based system can be and usually is split into two almost separate activities, i.e. design of the rule-base (declarative knowledge-base) and design and development of the inference engine;
- defining rules is in fact *declarative programming*; to certain degree it can be done independent of further inference mechanism and the way the rules will be used;
- development of the inference mechanism is based on *procedural programming*, separated from the knowledge-base development;
- declaratively encoded rules can be analyzed, verified, optimized, etc. as a kind of high-level, abstract data;
- the inference mechanism must work for different knowledge-bases; its design should be 'universal'.

Unlike writing an entire application all in a procedural language, rule-based applications contain a clearly distinguished knowledge-base and a module for interpreting the rules. The knowledge component can be exchanged, modified, analyzed, etc. without influencing the inference engine. The situation is a bit similar to Relational Database Systems — the data in the tables can be changed, it influences the results, but the procedures operating on the data (queries, transactions, build-in procedures) remain unchanged. In classical RDBS it is called *logical independence* and it follows from the famous Codd's postulates implemented in the ANSI-SPARC three level architecture [23]. In case of rule-based systems the situation is even more complex due to complex form of the rules expressed in specific logic-based languages.

A. Ligęza: *Logical Foundations for Rule-Based Systems*, Studies in Computational Intelligence (SCI) 11, 191–198 (2006)
www.springerlink.com

The declarative knowledge specified with rules constitutes a principal component responsible for the functional behavior of the system. As a separate module it should be designed so that the implemented system behaves in a desired way. In other words, this is the knowledge-base component which is responsible for accomplishing functional requirements and assures required services with satisfactory quality. As such, it should be validated, verified, tested and perhaps optimized. These activities will be roughly explained and issues of formal verification will be considered in details.

12.1 Validation, Verification, Testing and Optimization of Rule-Based Systems

The quality of software is an important focus in todays applications. Although there is no general agreement with respect to a unique, definite definition of software quality, the following features seem to be decisive as for assuring it [33]:

- **Reliability** — it is the capacity to assure the required level of functional services during the specified period of work.
- **Efficiency** — it is evaluation of the level of functional services with respect to required resources.
- **Functional Capability** — it is the capacity of satisfying user requirements with respect to the desired functions to be performed by the software.
- **Portability** — it is the capability of work in different environments.
- **Usability** — it is the easiness of using the software.
- **Maintenability** — it says how easy the system is to maintain.

In case of knowledge-based systems and especially rule-based components the above features give only a rough outline of what a quality knowledge-base should be like. For example, usability seems to be less important or even unimportant with respect to rule-based knowledge bases — it is user interface component which has decisive influence on that feature.

In case of rule-based systems the overall quality seems to be influenced by the following characteristic features:

- **Reliability** — the system should work and provide the required services at the required level.
- **Safety** — the system should not only work, but work in a safe way; this depends on elimination of potential failures and menaces on one hand, and ability to operate in case of certain faults.
- **Efficiency** — it should work in the best possible way at the minimal requirements of resources, e.g. memory, CPU time, etc.

The above features are not completely independent from one another; usually, assuring safety stays in opposite to efficiency. Reliability, on the other hand, is close to safety of operation[1].

Some further required features include also (at least to certain degree) *functional capability*, *portability* and *maintenability*. However, in case of rule-based systems these features are of somewhat specific nature.

For example, functional capability follows directly from the rule code since in fact designing rule-based systems is very much like writing executable specifications[2]. Achieving high functional capability is more the problem of the admitted knowledge specification language than the rules — if something can be expressed in the language, an appropriate rule can be added to the knowledge base.

Features as portability and maintenability on their turn, are achieved in a direct way due to the declarative programming approach which is intrinsic to rule-based systems design. They are also depending to certain degree on the knowledge specification language in use. Hence in fact, they are rather loosely related to the quality of the rule-base and the knowledge covered by it.

In the further part the interest will be focused on and around features such as *safety*, *reliability* and *efficiency* of rule bases and how these abstract, far-reaching aims can be translated into local, verifiable characteristics.

12.2 Verification: from General Requirements to Verifiable Characteristics

The expressive power of knowledge representation languages makes the scope of potential applications combined with modularity of RBS a very general and readily applicable mechanism. However, despite a vast spread-out in working systems, their theoretical analysis seems to constitute still an open issue with respect to analysis, design methodologies and verification of theoretical properties. Assuring *reliability*, *safety* and *efficiency* of rule-based systems requires both theoretical insight and development of practical tools. The general qualitative properties are translated into a number of more detailed characteristics defined in terms of logical conditions.

---

[1] The best analogy to clearly show the meaning and scope of these notions may be based on referring to an airplane: it is perfectly safe if it never starts — however, it is not reliable since it does not do its duty; obviously, it would also be inefficient. A reliable airplane performs according to desired schedule, it is safe if after any take-off there is exactly one landing, and it is efficient if the cost per passenger per one kilometer is lower than for other planes.

[2] One may say: 'You get whatever you want provided that you know how to ask for that in the accepted language'.

In fact, in order to assure safe and reliable performance, such systems should satisfy certain formal requirements, including completeness and consistency. To achieve a reasonable level of efficiency (quality of the knowledge-base) the set of rules must be designed in an appropriate way. Several theoretical properties of rule-based systems seem to be worth investigating, both to provide a deeper theoretical insight into the understanding of their capacities and assure their satisfactory performance, e.g. *reliability* and *safety* [3, 48, 101, 103, 107, 123]. Some most typical issues of theoretical verification include satisfaction of properties such as *consistency*, *completeness*, *determinism*, lack of *redundancy* or *subsumption*, etc. (see [3, 81, 101]). Several papers investigate these problems presenting particular approaches [25, 103, 107, 123]. A selection of tools is presented in [109]. Some modern approaches include [6, 49, 132].

The problems listed above become still more important in case of rule-based methodology applied to on-line, real-time control of dynamic systems (i.e. intelligent control, knowledge-based control) [48, 124], especially if safety issues are to be taken into account. Some of these problems may be of critical nature; for example, in case of lack of completeness, for certain states of the controlled system there are no rules to serve these states. This may make the system unreliable or unsafe.

A recent painful example of such lacks in safety[3] was the crash known as *The Warsaw Accident*, when a plane hit on high speed into an earth bank after successful landing under heavy weather conditions; switching to reverse thrust and opening spoilers (both for efficient breaking down) were disabled by an *intelligent control system.* For *certain* values of the speed of wheel spinning and weight *no action* was designed to be undertaken. It can be argued that static analysis of theoretical characteristics of such a system, in our case checking for completeness, could perhaps throw some light on missing rules identification and finding specification of gaps in the input state space served by the RBS applied for control.

The approach presented in this book is based on selection of verifiable characteristics (such as ones mentioned above) which are responsible for safety, reliability and efficiency. Then the initial qualitative requirements are translated to and expressed with requirements for satisfaction of such precisely defined characteristics. Each type of such characteristics is responsible for certain types of failures — hence, if one is able to check that the required characteristics are met, one may also be sure that the corresponding anomalies cannot occur. Below, taxonomies of such anomalies and characteristics are discussed.

[3] Caused in our rough and subjective interpretation by lack of completeness of the control algorithm; in fact the detailed analysis is more complex [119].

12.3 Taxonomies of Verifiable Features

In this section a new, generalized taxonomy of the issues concerning knowledge verification is put forward. The taxonomy covers most of the specific problems considered by other Authors, but groups together similar problems by taking into account the approach to perform appropriate check and deal with a specific anomaly. Further, logical definitions of specific anomaly classes are provided.

12.3.1 Verification of RBS: a Short Review

In this section a short review of selected, most common anomalies is presented. A kind of review of the theoretical problems specified by the authors with respect to verification of RBS properties is provided. The discussion is based mainly on the following positions [3, 25, 81, 83, 101, 109, 110, 111, 112, 123]. However, contrary to typical presentations, the discussion is *functionality-oriented*, i.e. we start from top-level, abstract characteristics required to achieve. Further, practical experience and theoretical discussion as well as author's former papers are taken into account.

Some most general classification of theoretical issues which undergo theoretical analysis can be one referring to the degree of influence they may have on system performance, and ranging from problems of *efficiency* and *elegance* of knowledge representation (e.g. redundancy and subsumption) to certain *critical errors* inside the encoded knowledge (e.g. *incompleteness* and *inconsistency* of specified knowledge). This point of view would be especially important when the analysis is to provide confidence about *reliability* and *safety* of an on-line (or even real-time) KBS working in safety-critical environment. On the other hand, *reliability* and *safety* are always to be considered w.r.t. KBS and its environment considered as two factors having joint influence on each other.

Note that, it is not the KBS itself, but always together with the controlled system and its environment which makes danger, crush or failure to occur. Thus, considering safety problems should be based on more global analysis covering not only the software system but its potential interaction with the environment. For example, redundancy, which is normally considered harmless, may slow down operation of a real-time critical system and lead to some serious consequences since the output would be delayed; on the other hand, even inconsistent or incomplete system can work well for a long time (even for years), provided that no use of its knowledge leading to direct manifestation of inconsistency is done or the uncovered inputs do not occur.

On the other hand, theoretical issues considered from the point of pure KBS theory (such as correctness, incompleteness, inconsistency, consistency with reality) may appear 'mathematically elegant' when analyzed at the level of theoretical definitions, but may turn out to become less attractive and hard to formalize when it comes to practical applications and efficient analysis.

Especially purely logical approaches, e.g. ones based on automated theorem proving, would seem promising, but are hardly applicable for realistic systems. The same applies to software verification in general case, where proving consistency of final code with initial specification is a hard, tedious, usually not a realistic task. But even if performed successfully, it is not the end of the problems.

12.3.2 Functional Quality Assignment

The approach of this work is relatively simple and conservative, based on good experience and success of RDB systems. Simultaneously, we try to be constructive: an engineering approach is pursued. An attempt is made at presentation of working classification of general issues of interest with respect to theoretical analysis, each of them having relatively different origin and ways to detect and deal with it.

Whenever appropriate, an idea of trouble detection procedure is outlined and suggestions about potential solutions are given. Further, the nature of potential problems is explained is case of omitting the analysis of a particular problem. In order to stay close to both engineering intuition and potential practical solutions, the discussion turns around simplified form of KBS, i.e. tabular RBS as introduced in Chap. 8.

From the point of view of top-level, desired functional specification, the following five characteristics are put forward:

- *safety*, i.e. the design of the rule-based system should assure that nothing dangerous would ever happen;
- *reliability*, i.e. the system should work and achieve its goals, possibly under any external circumstances;
- *admissibility*, i.e. the system should provide only admissible decisions or conclusions and should satisfy any constraints imposed on it;
- *quality*, i.e. the system should satisfy certain standards, especially satisfy explicit and implicit standards and user requirements;
- *efficiency*, i.e. the system should work in possibly most efficient way (perhaps even optimal) and should be specified in an efficient way (e.g. with the use of minimal number of rules, in the simplest form, etc.).

These top level characteristics appear to be both pairwise dependent and perhaps inconsistent with one another, i.e. satisfaction of one may lead to violating another one. For example, a safe plain would be one which never flies, but such a plain would not be reliable, not to tell about efficiency. Further, unfortunately, these characteristics are hardly expressible with the use of a formal specification. Thus, instead, the approach pursued in this work is based on translating them into quite technical, static analysis of selected features which can be defined formally with the use of logical specification. Below, a proposal of a general taxonomy is presented.

12.4 A Taxonomy of Verifiable Characteristics

The technical classification of theoretical properties of databases (considered within the extended paradigm, i.e. as data templates or knowledge facts), tabular rule based system as well as first-order based ones should possibly cover the complete spectrum of potential deficiencies. At the same time, the structure of the taxonomy should reflect verification paradigms, i.e. features analyzed with the same or similar tools should be grouped together. A proposal of a new taxonomy constituting an attempt at satisfying these principles and kept as transparent as possible is presented below. The proposal is based on papers such as [3, 101, 109, 110, 111, 112] and on author's recent proposals [63, 65].

The proposed taxonomy is a hierarchical one (two-level), and functionally similar detailed features are grouped together. The classification is applicable both to facts representing unconditional knowledge and rules divided into preconditions (*LHS*) and conclusion (*RHS*). The issues concerning various anomalies can be grouped and presented as follows:

- **Redundancy:**
  - identical rules,
  - subsumed rules,
  - equivalent rules,
  - unusable rules (ones never fired).
- **Consistency:**
  - indeterminism, ambiguous rules,
  - conflict, ambivalent rules,
  - logical inconsistency.
- **Reduction:**
  - reduction of rules,
  - canonical reduction of rules,
  - specific reduction of rules,
  - elimination of unnecessary attributes.
- **Completeness:**
  - logical completeness,
  - specific (physical) completeness,
  - detection of incompleteness,
  - identification of missing rules.

Recall that the above taxonomy is considered in the context of simple tabular systems, i.e. no rule chaining problems are taken into account (such as chains leading to contradiction, potential loops, dead-end condition, unreachable conclusion, etc.). Such problems usually require more complex analysis (e.g. recursive analysis of potential chains of rules combined in our case with potential inputs occurring at any stage; in a worst case scenario this may be equivalent to simulation of *all* potentially possible executions of the system, and as such it may be computationally intractable[4].

The proposed classification is somewhat general, but it covers many detailed cases mentioned in the literature. For example, subsumption of rules cover some four sub-cases of *Redundancy in pairs of rules* [101]. The case of *unnecessary IF conditions*, as discussed in [81], is a specific case of rule reduction discussed in this paper. On the other hand, as mentioned above, some checks requiring recursive analysis are not considered here. Note that in case of simple, reactive, forward-chaining systems it may be necessary to apply the same rule (or a sequence of rules) many times in turn, until external event changes the input (for example, in a supervisory system waiting for a special event). Further, a 'circular' rule may in fact be equivalent to iteration which may be necessary to load a counter, etc. and finishes only after expected amount of repetitions. Thus, 'circular' rules are not necessarily considered harmful.

Let us briefly skim through the proposed taxonomy of anomalies. In the next chapters, we shall analyze every class of problems in a brief way and we shall point to the principal checking approach. If applicable, logical specification of the appropriate condition to be verified will be provided.

[4] Well, analysis of complete set of cases of execution can be performed for certain, not-too-large systems; this, in fact, is carried out by PROLOG execution mechanism, and was also implemented in several rule-based verification systems, such as CHECK [103] w.r.t. circular rules detection or COVER [110] for deficiency detection in backward chaining systems. Certainly, a system once checked in a complete mode can be considered reliable and therefore can be safely used in the future, perhaps in multiple copies. On the other hand, we believe that for more complex systems, such complete checking is rather infeasible, especially if the system incorporates a language equivalent to first-order logic with equality and interpreted functions, but can be dealt with by keeping dynamic track of executed rules and results obtained at subsequent stages of inference, i.e. it can be performed on-line, during work of the system. This, however, would require that appropriate procedures are built-in into the rule interpreter.

13

Analysis of Redundancy

13.1 Redundancy of Knowledge Representation

The first group of anomalies in knowledge bases refers to issues concerning *redundancy of knowledge representation*, i.e. whether and how the current representation is inefficient through including unnecessary components. This issue has mostly no influence on system correctness and its functionality and potential or observed behavior. However, it can slow down its operation, and become a source of problems during modification or extension of the knowledge base. Redundancy in knowledge-bases (rule-bases) should be avoided due to the same reasons as in the classical relational databases — redundancy is a potential source of inconsistency when knowledge is updated. Last but not least, redundant elements occupy memory and make any analysis more difficult.

A general technique for dealing with redundant knowledge is to detect and remove the redundant components. Let R be a formula denoting the logical representation of a rule-base. There can be two points of view on redundancy: the *logical redundancy* and the *operational (functional) redundancy.* Below the appropriate definitions are given.

Definition 84. *The knowledge base represented with R is* logically redundant *if there exists R' obtained from R by removing certain component r, and there is $R \models R'$ and $R' \models R$.*

In other words, R is redundant if there is a possibility to reduce its size by removing at least one component (e.g. a rule in case of rule-based systems) and R' obtained in this way is logically equivalent to the initial formula.

A simple illustration of logical redundancy is the case of identical rules. Consider a rule base

$$R = r_1 \wedge r_2 \wedge \ldots \wedge r_m$$

where $r_i = \phi_i \longrightarrow h_i$ are some rules for $i = 1, 2, \ldots, m$. Let there be two rules in R, say i and j, such that $\phi_i = \phi_j$ and $h_i = h_j$; obviously, r_i is identical

A. Ligęza: *Logical Foundations for Rule-Based Systems*, Studies in Computational Intelligence (SCI) 11, 199–206 (2006)
www.springerlink.com

to r_j. Further, one of the rules can be removed from R so that R' is obtained, and $R' \equiv R$.

Definition 85. *The knowledge base represented with R is* functionally redundant *(or* operationally redundant*) if there exists a component r in R such that a new knowledge base R' obtained from R by removing r behaves exactly as R, i.e. for any input knowledge fact base FB, there is $R(FB) \equiv R'(FB)$.*

In other words, a rule base R is operationally redundant, if there exists a rule r which can be removed from it without influencing its behavior. A simple illustration of operational redundancy is the case of unusable rules. If one is sure that rule $r \in R$ will never be used (fired), then r can be removed from R without modifying its functional capabilities (its work). A rule is unusable (impossible to fire) if in any situation at least one of its literals occurring in the precondition formula will never be satisfied. This is so in case of unique predicate symbols (used only in this rule and not in the state descriptions or other rules) or unique attribute in case of tabular systems.

Logical redundancy implies operational redundancy — a system which is logically redundant is also operationally redundant. The example of unusable rules (due to unique condition) shows that the vice-versa is not necessarily true.

As for redundancy, the following issues have been identified so far:

- identical rules,
- subsumed rules,
- equivalent rules,
- unusable rules (ones never fired).

It is obvious that repeated, *identical rules* should be eliminated, as well as *redundant, equivalent data templates or rules*; the latter may however be difficult to identify without a theory supporting the proof of equivalence. *Redundant rules* (not necessarily identical — see the further discussion) can be detected and removed, leaving no more than one copy for each rule.

The most interesting is the case of *subsumed, less general rules.* A rule of the form $\phi \longrightarrow h$ subsumes (is more general than, or is stronger than) a rule $\phi' \longrightarrow h'$ if and only if it offers the possibility to draw stronger conclusions from weaker prerequisites, i.e. iff $\phi' \models \phi$ and $h \models h'$. This case will be analyzed in detail using both logical and algebraic approaches. *Subsumed rules* also can be eliminated, which has no influence on logical inference capability. However, in certain cases leaving a subsumed, more specific rule in knowledge base may be purposeful, for example it may affect the conflict resolution mechanism and inference control strategy [81].

The case of equivalent rules is analogous to the one of subsumed rules and follows it logically. Two rules are equivalent if one of them subsumes the other and vice-versa. Hence subsumption appears as generic, core issue since it covers also identical rules and equivalent ones.

A somewhat different case is the one of *unusable rules (ones never fired)*. Let $r = \phi \longrightarrow h$ be a rule suspected that it will never be fired. Such rules can be discovered with the use of analysis of feasible input states.

There can be in general two cases:

1) the set of feasible input states is described with joint formula Δ; then one is able to prove that $\Delta \not\models \phi$, i.e. the preconditions of the formula are never satisfied;
2) the infeasible states are described with a joint formula Γ; then one is able to prove that $\phi \models \Gamma$, i.e. if the precondition formula is satisfied, then we are in an infeasible state.

In general, such proofs are hard to complete; this is so since there is no way to find formulae Δ or Γ in a constructive way or they are too complex.

In literature one can meet certain simplified criteria, such as occurrence of unique attribute or predicate symbol in preconditions of the rule.

Finally, when analyzing redundancy, there can be also the case of redundancy within the preconditions of a rule. A typical such case happens if some attributes are unnecessary. The case of *unnecessary attributes* (or *too many attributes*), is not necessarily easy to identify — semantic, domain dependent knowledge is necessary. Such a case is the consequence of existence of functional dependencies among attributes, and since it is relatively well studied in the theory of RDB, it will not be considered here.

13.2 Subsumption

Let us consider the most general case of subsumption; some particular definitions are considered in [3, 103, 123].

A rule subsumes another rule if the following conditions hold:

- the precondition part of the subsuming rule is *weaker* (more general) than the precondition of the subsumed rule (the subsuming rule succeeds in more situations than the subsumed one),
- the conclusion part of the subsuming rule is *stronger* (more specific) than the conclusion of the subsumed rule (the subsuming rule provides more information than the subsumed one).

Particular instances of subsumption follow from holding either the first or the second condition, while keeping equivalence or identity of the other parts of the rules. The case of redundant rules [3, 103, 123], i.e. when one of them succeeds in the same situation as another rule and both the rules have the same conclusions can be considered as the simplest case of subsumption.

Let the rules r and r', satisfy the following assumption: $\phi' \models \phi$ and $h \models h'$. The subsumed rule can be eliminated according to the following scheme

$$\frac{\begin{array}{l} r: \phi \longrightarrow h \\ r': \phi' \longrightarrow h' \end{array}}{r: \phi \longrightarrow h}.$$

For intuition, a subsumed rule can be eliminated because it produces weaker results and requires stronger conditions to be satisfied; thus any of such results can be produced with the subsuming rule.

13.2.1 Subsumption in First Order Logic

In order to detect subsumption in First-Order Logic one has to employ general theorem proving procedures. For example, in case of rules having complex precondition expressed in the DNF, backward dual resolution can be applied in a straightforward way [53]. However, employing general proof procedure for subsumption analysis is rather of theoretical interest than of practical importance.

In case of simple rules having preconditions in form of simple conjunctive formulae, using pattern matching algorithms based on unification seems to be satisfactory.

Consider the following simple example which is aimed at illustrating the idea. Let there be given two rules r_1 and r_2, where:

$$r_1 \colon p(X) \wedge q(b) \longrightarrow h(X) \wedge g(b),$$
$$r_2 \colon p(a) \wedge q(b) \wedge s(Y) \longrightarrow h(a).$$

Now, for substitution $\sigma = \{X/a\}$ we have $LHS(r_1)\sigma \subseteq LHS(r_2)$. Hence $LHS(r_2) \models LHS(r_1)$ — the preconditions of r_2 are in fact more detailed. Simultaneously, $RHS(r_2) \subseteq RHS(r_1\sigma)$ and hence $RHS(r_1) \models RHS(r_1)$ — the conclusion of r_1 is stronger. Hence rule r_2 is subsumed by rule r_1. In fact, having defined rule r_1 , r_2 is no longer necessary.

13.2.2 Subsumption in Tabular Systems

Consider now the case of simple tabular systems encoding rules in attributive logic. Consider two rules, r and r'. The condition for subsumption in case of tabular rule format takes the algebraic form $t'_j \subseteq t_j$, for $j = 1, 2, \ldots, n$ and $h' \subseteq h$. If it holds, then rule r' can be eliminated according to the following scheme:

rule	A_1	A_2	...	A_j	...	A_n	H
r	t_1	t_2	...	t_j	...	t_n	h
r'	t'_1	t'_2	...	t'_j	...	t'_n	h'

rule	A_1	A_2	...	A_j	...	A_n	H
r	t_1	t_2	...	t_j	...	t_n	h

For example, in the following tabular system the first rule subsumes the second one:

rule	A_1	A_2	A_3	A_4	H
r	7	$[2,9]$	$[3,5]$	$\{r,g,b\}$	$\{a,b,c\}$
r'	7	$[3,5]$	4	$\{b,r\}$	$\{a,c\}$

The check is based on pure algebraic test specified with rule (3.2) presented in Sect. 3.6. In general case, the check for subsumption may require $m(m-1)/2$ comparisons among rules (where m is the number of rules in the analyzed table), however most of them will fail quickly. In large tables, the rules may be further structured w.r.t partition of certain attributes, so that subsumption may be checked only within smaller subgroups of rules.

13.3 Verification of Subsumption in XTT — a Prolog Code

Consider an example PROLOG code for verification of subsumption of attributive rules encoded in the form of an XTT table.

The rule format is presented below.

```
% f(<attribute_name>,<value_type>,<value>)
% <value_type>: atomic, set, interval, ...

% rule format:
% rule(<table_number>,
%       <rule_number>,
%       [<precondition_list>],
%       [<retract_list>],
%       [<assert_list>],
%       [<decision_list>],
%       <next_table>,
%       <next_rule in next_table>,
%     ).
```

For practical examples of rule encoding see the Thermostat example presented in Chap. 20.

The following facts are declared as dynamic ones.

```
:- dynamic f/3.
:- dynamic rule/10.
:- dynamic set/2.
```

The following predicate *vsu/1* performs verification of subsumption over a table being its argument.

```
vsu(T):-
        rule(T,N1,P1,R1,A1,D1,_,_),
        rule(T,N2,P2,R2,A2,D2,_,_),
        N1 \= N2,
```

```
        subsumes(P1,P2),
        covers(R1,R2),
        covers(A1,A2),
        covers(D1,D2),
        write('*** Rule: '),
        write(T),write('.'),write(N1), write(' subsumes rule: '),
        write(T),write('.'),write(N2), nl, fail.

vsu(T):-
        write(' No more subsumption of rules in table '), write(T), nl.
```

The idea of the check is straightforward. The predicate *vsu/1* reads pairs of rules from table T and compares them. Subsumption holds iff preconditions $P1$ of some rule subsume preconditions $P2$ of another rule ($subsumes(P1, P2)$), and simultaneously all the results of the first rule are stronger than the ones of the other rule ($covers(R1, R2)$, $covers(A1, A2)$, $covers(D1, D2)$). The main predicate is *subsumes/2* for verification of subsumption amongst the preconditions of rules, and *covers/2* for verification of subsumption amongst the conclusions, including the retract, assert and output components. They are defined as follows.

Here are the definitions of auxiliary predicates:

```
subsumes([],_):- !.
subsumes([Fact|Facts],L):- sub(Fact,L), subsumes(Facts,L).

sub(Fact,L):-
        retractall(f(_,_,_)),
        assertlist(L),
        valid(Fact).
assertlist([]):- !.
assertlist([F|R]):- assert(F), assertlist(R).

covers(_,[]):- !.
covers(L,[Fact|Facts]):- cov(L,Fact), covers(L,Facts).

cov(L,Fact):-
        retractall(f(_,_,_)),
        assert(Fact),
        member(F,L),
        valid(F).
```

In order to verify subsumption among preconditions one has to check if any fact of the more general list ($P1$) is valid provided that the less general list ($P2$) is valid; the check consists of asserting the list into the global memory and subsequent recursive verification of all the facts of $P1$.

In order to verify that conclusions of the first rule cover the ones of the second rule one has to check that the stronger conclusions cover any fact of the conclusions of the second rule. The check is analogous, but this time the fact to be covered is asserted to the global knowledge base and a fact covering it is sought in the covering list. This is so in order to use the same predicate *valid/1* which describes various possibilities of matching a fact against a fact base in the memory; the difference follows from the interpretation of facts in preconditions as *internal disjunction* and in conclusions as *internal conjunction* (see Sect. 3.5). The *valid/1* predicate takes as its argument a single fact (normally from the list of preconditions) and checks if it is satisfied in view of the global fact base. It is defined as follows.

```
valid(f(A,atomic,V)) :- f(A,atomic,V),!.
valid(f(A,natomic,V)) :- f(A,atomic,W), V \== W, !.
valid(f(A,set,Set)) :-
        f(A,atomic,V),
        set(Set,SetValue),
        member(V,SetValue),!.
valid(f(A,set,Set)) :-
        f(A,set,SA),
        set(SA,SAValue),
        set(Set,SetValue),
        subset(SAValue,SetValue),!.
valid(f(A,nset,Set)) :-
        f(A,atomic,V),
        set(Set,SetValue),
        \+ member(V,SetValue),!.
valid(f(A,nset,Set)) :-
        f(A,set,SA),
        set(SA,SAValue),
        set(Set,SetValue),
        intersection(SAValue,SetValue,[]),!.
valid(f(A,interval,i(B,E))) :-
        f(A,atomic,V),
        V >= B,
        V =< E,!.
valid(f(A,ninterval,i(B,_))) :-
        f(A,atomic,V),
        V < B,!.
valid(f(A,ninterval,i(_,E))) :-
        f(A,atomic,V),
        V > E,!.
valid(f(A,interval,i(B,E))) :-
        f(A,interval,i(BB,EE)),
        BB >= B,
        EE =< E,!.
valid(f(A,ninterval,i(_,E))) :-
        f(A,interval,i(BB,_)),
        E < BB,!.
valid(f(A,ninterval,i(B,_))) :-
        f(A,interval,i(_,EE)),
        EE < B,!.
```

The precise definition of *valid/1* must be adjusted to the detailed form of the attributive language in use. The provided code allows for use of individual values, sets, intervals, also together with negated relation (*natomic, nset*, and *ninterval* having the meaning of $\neq$, $\notin$ (set) and $\notin$ (interval) respectively.

14

Analysis of Indeterminism and Inconsistency

14.1 Indeterminism and Inconsistency of Rules

Problems of indeterminism or ambiguity, ambivalence and inconsistency of rules refer to the generic topic of internal consistency of the rule-base. This is the case when consistent application of the rules may lead to ambiguous or inconsistent results. In the simplest case, different results may be inferred when different inference control strategies are applied. In the most serious case the inferred results may be logically inconsistent.

In general, one can consider the following cases of anomalies:

- indeterminism, ambiguous rules;
- conflict, ambivalent rules;
- logical inconsistency.

The lack of *determinism or uniqueness* may lead to *ambiguous results. Ambiguous results* may take place in case when two (or more) rules can be applied for the same input, but their outputs are different. Consider two rules r_1 and r_2 belonging to the same rule-base, where r_1 is of the form $\psi_1 \longrightarrow h_1$ and r_2 is of the form $\psi_2 \longrightarrow h_2$.

Definition 86. *Two rules r_1 and r_2 are* ambiguous *or form* indeterministic *set of rules if there exists a state described by formula ϕ, such that simultaneously $\phi \models \psi_1$ and $\phi \models \psi_2$ and $h_1 \neq h_2$.*

In other words, ambiguous rules are ones which can be simultaneously fired, but their conclusions are different.

It may be the case that such a result is harmless, or even intended, but in case of reactive control systems the situation like that should be carefully analyzed. From logical point of view, rules as above are potentially ambiguous iff $\psi_1 \wedge \psi_2$ is a satisfiable formula. In this case one can say that *logical* or potential indeterminism exists. If there exists a physical state for which the rules can be fired, one can say that the rules are *physically* or practically ambiguous.

A. Ligêza: *Logical Foundations for Rule-Based Systems*, Studies in Computational Intelligence (SCI) **11**, 207–211 (2006)
www.springerlink.com

A more dangerous case is the one of *conflicting, ambivalent rules*, i.e. when the simultaneously produced outputs both cannot be correct with respect to the intended interpretation I (external world).

Definition 87. *Two rules r_1 and r_2 are* conflicting *or form* ambivalent *set of rules if there exists a state described by formula ϕ, such that simultaneously $\phi \models \psi_1$ and $\phi \models \psi_2$ but $\not\models_I h_1 \wedge h_2$ under the assumed interpretation I.*

In other words, ambivalent (conflicting) rules can be simultaneously fired, but their conclusions are in conflict — h_1 and h_2 cannot be simultaneously true.

For example, there exist many devices which can be in one and only one state at a certain instant of time, and concluding that such a device takes simultaneously two different states leads to physical inconsistency. For example, all the bistable elements, such as switches or relays, can be either *on* or *off*. In the above case two ambivalent rules would be conflicting if apart from overlapping preconditions also the formula $h \wedge h'$ would be unsatisfiable under the assumed interpretation.

The conflict may become *logical inconsistency* if certain two conclusions are logically inconsistent, either in direct case (if one is the negation of the other), or in an indirect case (when assuming that both are simultaneously true allows for formal demonstration that there is logical inconsistency).

Definition 88. *Two rules r_1 and r_2 are* inconsistent *if there exists a state described by formula ϕ, such that simultaneously $\phi \models \psi_1$ and $\phi \models \psi_2$ but $\not\models h_1 \wedge h_2$.*

In other words, ambivalent (conflicting) rules can be simultaneously fired, but their conclusions are logically inconsistent; for example $h_1 = \neg h_2$ which cannot be true under any interpretation.

For obvious reasons such problems should be detected and carefully analyzed.

Note that a common characteristic and partially the source of the above problems follows from *overlapping* preconditions of the rules. Hence, the basic verification procedure should be able to discover any pair of rules such that their preconditions can be simultaneously satisfied.

From logical point of view, rules are potentially overlapping (with respect to their preconditions) iff $\psi_1 \wedge \psi_2$ is a satisfiable formula. In this case one can say that *logical* or potential overlapping is observed. If there exists a physical state for which the rules are both fire-able, one can say that the rules are *physically* or practically overlapping.

14.2 Consistency Analysis

In this section the problem of determinism and two following issues, i.e. the one of conflict and inconsistency are discussed.

14.2.1 Determinism

A set of rules is *deterministic* iff no two different rules can succeed for the same state. A set of rules which is not deterministic is also referred to as *ambiguous.*

The idea of having a deterministic system is based on a priori elimination of 'overlapping' rules, i.e. ones which operate on a common situation. The aim of analysis is obvious — to detect (distinguish) the case of two or more rules applicable the same situation.

Consider the following two rules:

$$r_1 \colon \psi_1 \longrightarrow h_1,$$
$$r_2 \colon \psi_2 \longrightarrow h_2.$$

From purely logical point of view the system is deterministic iff the conjunction of the precondition formulae $\psi_1 \wedge \psi_2$ is unsatisfiable. For certain technical systems it is enough to check that such a conjunction is unsatisfiable under a specific interpretation referring to the domain of interest [53]. A typical example is of the form $\psi_1 = switch(on)$ and $\psi_2 = switch(off)$, and obviously the formula $\psi_1 \wedge \psi_2$ is false under the intended interpretation — the *switch* can be either *on* or *off.*

Consider now the case of attribute knowledge representation, as below

rule	A_1	A_2	$\ldots$	A_j	$\ldots$	A_n	H
r_1	$t_{1,1}$	$t_{1,2}$	$\ldots$	$t_{1,j}$	$\ldots$	$t_{1,n}$	h_1
r_2	$t_{2,1}$	$t_{2,2}$	$\ldots$	$t_{2,j}$	$\ldots$	$t_{2,n}$	h_2

Calculation of $\psi_1 \wedge \psi_2$ is straightforward: for any attribute A_j there is an atom of the form $A_j = t_{1,j}$ in ψ_1 and $A_j = t_{2,j}$ in ψ_2, $i = 1, 2, \ldots, n$. Now, one has to find the intersection of $t_{1,j}$ and $t_{2,j}$ — if at least one of them is empty (e.g. two different values; more generally $t_{1,j} \cap t_{2,j} = \emptyset$), then the rules are disjoint. The check is to be performed for any pair of rules.

14.2.2 Conflict and Inconsistency

From practical point of view deterministic systems are easier for implementation. In case of indeterministic system there may be the case that two or more rules are simultaneously applicable. It is the problem then of the so-called *conflict resolution mechanism* to select a single rule to be fired. Note that if a system is deterministic, no conflict resolution mechanism is necessary. On the other hand, in certain systems indeterminism is inherent in the set of rules, while conflict situations are to be solved with appropriate inference control mechanism.

In design of knowledge rule-based systems one can encounter further theoretical problems; two most important ones following from the lack of determinism are as follows (see also [103, 123]):

- **Conflict** — two (or more) rules are applicable to the same input situation but the results are conflicting (under the assumed interpretation).
- **Inconsistency** — here understood as logical inconsistency (unsatisfiability under any interpretation).

Note that problems of *conflicting* and *inconsistent* rules [103, 123] are specific cases of indeterminism. In tabular systems with no explicit negation purely logical inconsistency cannot occur; it always follows from the intended interpretation and thus it falls into the class of conflicts.

The case of conflicting rules is a potential source of errors, e.g. ambivalent, and therefore unpredictable behavior. Conflicting rules, however, are a subclass of indeterministic ones; thus, any pair of such rules will be eventually discovered after checking for determinism; they should be further analyzed by domain experts. This is so, since different conclusions of two overlapping rules do not necessarily mean 'real conflict', i.e. both suggested solutions can be admissible. This is the typical case of *Decision Support Systems* suggesting one or several solutions valid for certain situation. Depending on the inference strategy, either one of them (selected in an arbitrary way or according to some predefined strategy) or even all of them can be applied. On the other hand, a real conflict exists usually in case of unique decision to be undertaken, e.g. if some resources are indivisible or some variables can be assigned a unique value only.

14.3 Verification of Indeterminism: a Prolog Code

Consider an example PROLOG code for verification of indeterminism of attributive rules encoded in the form of an XTT table. The rule format is as presented before. For practical examples of rule encoding see the Thermostat example presented in Chapter 20.

The main verification predicate *vnd/1* takes a table of rules as its argument; the core of its code is shown below.

```
vnd(T):-
       rule(T,N1,P1,_,_,_,_,_),
       rule(T,N2,P2,_,_,_,_,_),
       N1 \== N2,
       overlaps(P1,P2),
       write('*** Rule: '),
       write(T),write('.'),write(N1), write(' overlaps with rule: '),
       write(T),write('.'),write(N2), nl, fail.
vnd(T):-
       write('No more overlapping of rules in table '), write(T), nl.
```

The check is accomplished by verifying if the preconditions of the rules overlaps. This is done recursively by the predicate *overlaps/2*. The different alternative possibilities of overlapping for two atomic formulae are defined with *over/2* predicate.

```
overlaps([],[]):- !. overlaps([F|Facts],[G|Gacts]):-
over(F,G),overlaps(Facts,Gacts).

over(f(A,atomic,V),f(A,atomic,V)):- !.
over(f(A,natomic,_),f(A,natomic,_)):- !.
over(f(A,natomic,V),f(A,atomic,W)):- V \== W,!.
over(f(A,atomic,V),f(A,natomic,W)):- V \== W,!.
over(f(A,atomic,V),f(A,set,S)):-
     set(S,SetValue), member(V,SetValue), !.
over(f(A,set,S),f(A,atomic,V)):-
     set(S,SetValue), member(V,SetValue), !.
over(f(A,set,S),f(A,set,Q)):-
     set(S,SetValue), set(Q,QetValue),
     member(V,SetValue), member(V,QetValue), !.
over(f(A,atomic,V),f(A,interval,i(B,E))):- B =< V, V =< E,!.
over(f(A,interval,i(B,E)),f(A,atomic,V)):- B =< V, V =< E,!.
over(f(A,interval,i(B,E)),f(A,interval,i(BB,EE))):-
     BB =< E, B =< EE,!.
over(f(A,interval,i(BB,EE)),f(A,interval,i(B,E))):-
     BB =< E, B =< EE,!.
```

The rules with overlapping preconditions are listed for further analysis. If definitions of conflicting conclusions can be specified in a constructive way (e.g. by enumeration of physically inconsistent cases), the above procedure can be extended to perform the appropriate check on the conclusions of overlapping rules as well.

15

Reduction of Rule-Based Systems

Problems of minimal or maybe optimal representation refer to the possibility of transformation of the initial knowledge table to some, possibly simplest, form, i.e. simplification. This can be obtained through reduction of the table. The reduced form, should, however, be logically equivalent to the input table.

Note that reduction understood in this way is different from simple *elimination* of unnecessary rules (e.g. subsumed or equivalent ones). In looking for minimal representation of rules one can *glue* two or more rules so that a single, more general rule is obtained. The gluing operation is the basic mechanism for reduction.

The most important issues specific for finding a minimal representation for a given set of rules are the following:

- reduction of rules to maximally reduced form,
- partial reduction of rules to canonical form,
- specific reduction of rules,
- elimination of unnecessary attributes.

Reduction of rule base to maximally reduced forms means replacing two or more items with a single, equivalent rule by an operation resembling 'gluing' the preconditions of them. Note that from logical point of view all of the activities referring to cleaning-up most of the anomalies do lead to logically equivalent knowledge base, but a simpler (more elegant) one.

The reduction can be total (i.e. maximal; no further reduction is possible) or partial. Such *partial reduction of data templates or rules* may lead to a specific, unique form, called *canonical form* [64][1]. It can be noticed that the result of total reduction may be, in general case, not unique. It may be wise then, to stop at a certain stage of partial reduction, but such that the result would be defined in a unique way. More on that can be found in [64].

[1] For intuition, a canonical form of a tabular system satisfies the condition that the values appearing in any column do not overlap or are identical; the canonical form may be unique for a specific tabular system [72–74].

A. Ligęza: *Logical Foundations for Rule-Based Systems*, Studies in Computational Intelligence (SCI) **11**, 213–218 (2006)
www.springerlink.com

Specific partial reduction of data templates or rules is a kind of reduction where two or more values, terms or formulae are replaced with another, more general one. Normally, reduction leads to elimination of unnecessary terms or formulae; in case of specific partial reduction the term or formula 'survives' but it is generalized w.r.t. the original one.

Elimination of unnecessary attributes may be performed either with the use of semantic knowledge about functional dependencies or through total reduction (if applicable). In such a case the output does not appear to depend on some attribute(s), which disappear during the reduction process.

15.1 Generation of Minimal Forms of Tabular Rule-Based Systems

Minimization of knowledge representation can be achieved by elimination of redundant and subsumed rules and by appropriate joining together selected rules which are in certain sense 'complementary'. The main idea of the second case is based on the principles of backward dual resolution and is referred to as *reduction of rules.*

15.1.1 Total and Partial Reduction

Reduction of rules is an operation similar to finding minimal representation for propositional calculus formulae or boolean combinatorial circuits. The main idea of reduction of rules is to minimize the number of rules without influencing the potential capabilities of the system for inferring new knowledge. Efficient reduction may be accomplished by replacing a number of rules having the same conclusions with a single equivalent rule.

Consider k rules with the same conclusion, such that their preconditions differ only with respect to ω_i, $i = 1, 2, \ldots, k$, where $\omega_i = (A_j \in t_{ij})$ defines the value of the same single attribute A_j. Assume that the following completeness condition holds

$$\models \omega_1 \vee \omega_2 \vee \ldots \vee \omega_k.$$

With respect to the dual resolution principle the following reduction scheme can be applied:

$$\frac{\begin{array}{llcl} r^1: & \phi \wedge \omega_1 & \longrightarrow & h \\ r^2: & \phi \wedge \omega_2 & \longrightarrow & h \\ \vdots & & & \\ r^k: & \phi \wedge \omega_k & \longrightarrow & h \end{array}}{\begin{array}{llcl} r: & \phi & \longrightarrow & h \end{array}}.$$

For intuition, the preconditions of the formulae are replaced by a joint condition representing the disjunction of them; roughly speaking, the sets described with the preconditions are 'glued' together into a single set. The resulting rule is logically equivalent to the set of initial rules.

Using the tabular knowledge representation, reduction takes the following form:

rule	A_1	A_2	...	A_j	...	A_n	H
r^1	t_1	t_2	...	t_{1j}	...	t_n	h
r^2	t_1	t_2	...	t_{2j}	...	t_n	h
$\vdots$	$\vdots$	$\vdots$		$\vdots$		$\vdots$	$\vdots$
r^k	t_1	t_2	...	t_{kj}	...	t_n	h

rule	A_1	A_2	...	A_j	...	A_n	H
r	t_1	t_2	...	-	...	t_n	h

provided that $t_{1j} \cup t_{2j} \cup \ldots \cup t_{kj} = D_j$. Of course, the rules $r^1, r^2, \ldots, r^k$ are just some selected rows of the original table containing all the rules. If $\mathbf{B}$ denotes the original table, its (maximally) reduced form will be denoted as $Red(\mathbf{B}) = \mathbf{B}^*$.

Note that a strong requirement for reduction follows from the fact that the values of attributes other than the reduced ones must be identical. This is important if one insists on preserving logical equivalence between the initial and generated table. In some cases, however, it may be of interest to produce a new rule more general with respect to the selected attribute (i.e. making use of gluing its partial values), while admitting a 'slight' restriction concerning the rest of the preconditions. The following form of reduction can also be proposed:

$$\frac{\begin{array}{lll} r^1: & \phi^1 \wedge \omega_1 & \longrightarrow h \\ r^2: & \phi^2 \wedge \omega_2 & \longrightarrow h \\ \vdots & & \\ r^k: & \phi^k \wedge \omega_k & \longrightarrow h \end{array}}{\begin{array}{lll} r: & \phi^1 \wedge \phi^2 \wedge \ldots \wedge \phi^k & \longrightarrow h \end{array}}.$$

This may be reasonable provided that formula $\phi^1 \wedge \phi^2 \wedge \ldots \wedge \phi^k$ is useful and not too restrictive, and may be applied in completeness verification. Note that full logical equivalence may be no longer preserved, however, disjunction of the preconditions of mother rules follows from the precondition of the reduced formula. If the resulting reduced rule can be applied, then at least one of the rules above can be applied as well, and the produced conclusion is of course identical.

In general, the reduction can be total (maximal) or partial, depending on the needs. The total reduction may result in several, different results, sometimes hard to compare. Partial reduction can be stopped at certain point, e.g. when the canonical form is generated [64], which allows for easy comparison of results and further algebraic operations.

15.1.2 Specific Partial Reduction

In certain cases a complete reduction as shown above may turn out to be inapplicable; however, it may still be possible to simplify the set of rules if only the sub-formulae ω_i, $i = 1, 2, \ldots, k$ can be replaced with a single equivalent formula. In our case, a collection of certain elements can be always replaced by a subset containing all of them (and nothing more), while a collection of intervals can be replaced with their sum (which may be a single, convex interval). In general, let us assume that $\omega_1 \vee \omega_2 \vee \ldots \vee \omega_k \models \omega$, and $\omega \models \omega_1 \vee \omega_2 \vee \ldots \vee \omega_k$. The reduction can take the following logical form:

$$\frac{\begin{array}{lll} r^1: & \phi \wedge \omega_1 & \longrightarrow h \\ r^2: & \phi \wedge \omega_2 & \longrightarrow h \\ \vdots & & \\ r^k: & \phi \wedge \omega_k & \longrightarrow h \end{array}}{\begin{array}{lll} r: & \phi \wedge \omega & \longrightarrow h \end{array}}.$$

Formula ω must be expressible within the accepted language. In case of a single attribute the internal disjunction can be applied just by specifying the appropriate subset.

Using the tabular knowledge representation, partial reduction takes the following form:

rule	A_1	A_2	...	A_j	...	A_n	H
r^1	t_1	t_2	...	t_{1j}	...	t_n	h
r^2	t_1	t_2	...	t_{2j}	...	t_n	h
$\vdots$	$\vdots$	$\vdots$		$\vdots$		$\vdots$	$\vdots$
r^k	t_1	t_2	...	t_{kj}	...	t_n	h

rule	A_1	A_2	...	A_j	...	A_n	H
r	t_1	t_2	...	t	...	t_n	h

provided that $t_{1j} \cup t_{2j} \cup \ldots \cup t_{kj} = t$. As above, the rules $r^1, r^2, \ldots, r^k$ are just some selected rows of the original table containing all the rules. Moreover, we extend the notation on partial reduction as well, i.e. if **B** denotes the original table, its (maximally) reduced form using also partial reduction whenever possible will be denoted as $Red(\mathbf{B}) = \mathbf{B}^*$ (or $Red(\mathbf{\Phi}) = \mathbf{\Phi}^*$).

Note also that in the case of partial reduction generation of more restrictive formulae is possible in a way analogous to the case of pure reduction. The transformation takes the form:

$$\frac{\begin{array}{lll} r^1: & \phi^1 \wedge \omega_1 & \longrightarrow h \\ r^2: & \phi^2 \wedge \omega_2 & \longrightarrow h \\ \vdots & & \\ r^k: & \phi^k \wedge \omega_k & \longrightarrow h \end{array}}{\begin{array}{lll} r: & \phi^1 \wedge \phi^2 \wedge \ldots \wedge \phi^k \wedge \omega & \longrightarrow h \end{array}}$$

and it may be reasonable if the formula $\phi^1 \wedge \phi^2 \wedge \ldots \wedge \phi^k \wedge \omega$ is not too restrictive. The idea of canonical form applies as well to specific partial reduction.

15.2 Reduction of Tabular Systems — a Prolog Code Example

Consider an example PROLOG code for computer-aided reduction of attributive rules encoded in the form of an XTT table. The pairs of rules that can be reduced are listed as an output. The main predicate for checking reduction possibility is *vpr/1*.

The rule format was presented before.

```
vpr(T):-rule(T,N1,P1,R1,A1,D1,_,_),
        rule(T,N2,P2,R2,A2,D2,_,_),
        N1 \== N2,
        reduce(P1,P2,F),
        R1==R2,A1==A2,D1==D2,
        write('*** Rule: '),
        write(T),write('.'),write(N1),
        write(' may be glued  with rule: '),
        write(T),write('.'),write(N2),
        write('  reduced fact: '), write(F), nl, fail.
vpr(T):-
        write('No more reduction of rules in table '),
        write(T), nl.
```

The main reduction predicate is *reduce/3* which takes as arguments lists of preconditions of the rules to be glued and produces a fact being the result of gluing certain two facts selected from the preconditions lists.

Due to different types of facts there are different possibilities of reduction — they are specified below with the *red/3* predicate.

```
reduce([],[]):-!.
reduce(P1,P2,F):-
       select(F1,P1,PF1),
       select(F2,P2,PF2),
       PF1==PF2,
       red(F1,F2,F).

red(f(A,atomic,V),f(A,atomic,U),F):-
       \+number(V),\+number(U),F=f(A,set,[V,U]),!.
red(f(A,atomic,V),f(A,set,S),F):-
       set(S,SetValue), F=f(A,set,[V|SetValue]),!.
red(f(A,set,S),f(A,atomic,V),F):-
       set(S,SetValue), F=f(A,set,[V|SetValue]),!.
red(f(A,set,S),f(A,set,Q),F):-
       set(S,SetValue), set(Q,QetValue),
```

```
        union(SetValue,QetValue,UnionValue),
        F=f(A,set,UnionValue),!.
red(f(A,atomic,V),f(A,interval,i(B,E)),F):-
        number(V),V >= B, V=< E, F=f(A,interval,i(B,E)),!.
red(f(A,interval,i(B,E)),f(A,atomic,V),F):-
        number(V),V >= B, V=< E, F=f(A,interval,i(B,E)),!.
red(f(A,interval,i(B,E)),f(A,interval,i(BB,EE)),F):-
        BB =< E, E =< EE, BBB=min(B,BB),EEE=max(E,EE),
        F=f(A,interval,i(BBB,EEE)),!.
red(f(A,interval,i(B,E)),f(A,interval,i(BB,EE)),F):-
        B =< EE, EE =< E, BBB=min(B,BB),EEE=max(E,EE),
        F=f(A,interval,i(BBB,EEE)),!.
```

The above code finds pairs of rules which are possible to be glued and produces the glued fact; the other elements of the lists are kept unchanged. This code can be combined with subsumption checking for elimination of less general rules as well.

16

Analysis of Completeness

16.1 Completeness of Rules

The problem of completeness verification can be defined as checking if all possible inputs are served by at least one rule. Practically, this means that for any combination of input values and conditions, preconditions of at least one rule should be satisfied. If this is not the case, the system is in certain sense incomplete. This may be a desired effect, but in most cases of practical dynamic systems (control, supervisory or decision support ones) incompleteness means a design error — there is a gap in knowledge.

The most important issues concerning analysis of completeness include the following ones:

- verification of logical completeness,
- verification of specific (physical) completeness,
- detection of incompleteness,
- identification of missing rules.

Apart from inconsistency analysis, verification of completeness seems to be one of the most important issues for designing safe and reliable systems.

Logical (total) completeness means that the disjunction of preconditions of all the rules form a tautology, i.e. no matter what input combination occurs, it will be served.

Consider a set of rules of the form:

$$\begin{array}{lll} r_1: & \phi_1 & \longrightarrow & h_1, \\ r_2: & \phi_2 & \longrightarrow & h_2, \\ \vdots & & & \\ r_m: & \phi_m & \longrightarrow & h_m. \end{array}$$

Definition 89. *A set of rules $R = \{r_1, r_2, \ldots, r_m\}$ is* logically complete *iff*

$$\models \phi_1 \vee \phi_2 \vee \ldots \vee \phi_m,$$

i.e. if the joint precondition formula is a tautology.

A. Ligęza: *Logical Foundations for Rule-Based Systems*, Studies in Computational Intelligence (SCI) 11, 219–228 (2006)
www.springerlink.com

Logical completeness means that whatever the input is, preconditions of at least one rule are satisfied. So at least one rule can be fired.

In practice, not all input states may be admissible, or the system may be designed to work only for certain inputs (in a specific *context*). For most of practical systems logical completeness constitutes a requirement which is too strong. *Specific (partial, physical) completeness* means that the scope of inputs such that the system is capable of dealing with is explicitly defined with a formula (restricting conditions) specifying the admissible input space.

Definition 90. *A set of rules* $R = \{r_1, r_2, \ldots, r_m\}$ *is* specifically complete *(physically complete) with respect to context defining formula* Ψ *iff*

$$\Psi \models \phi_1 \vee \phi_2 \vee \ldots \vee \phi_m,$$

i.e. if the joint precondition formula is satisfied within the predefined context Ψ.

Specific completeness is close to practical understanding of completeness. If all the physical input states that should be served with the rules are defined with formula Ψ, then specific completeness with regard to Ψ means also physical completeness with respect to all the states satisfying Ψ.

In both cases, if the system does not satisfy completeness requirements, it may be of interest to determine gaps in the system input which is served by the rules, i.e. to generate a specification of unserved inputs; this means that *detection of incompleteness* and subsequent *identification of missing rules* should be carried out. Logically, one would have to find formulae $\phi'_1, \phi'_2, \ldots, \phi'_k$, such that $\models \phi_1 \vee \phi_2 \vee \ldots \vee \phi_m \vee \phi'_1 \vee \phi'_2 \vee \ldots \vee \phi'_k$, but such that also $(\phi_1 \vee \phi_2 \vee \ldots \vee \phi_k) \wedge (\phi'_1 \vee \phi'_2 \vee \ldots \vee \phi'_m)$ is never satisfied.

16.2 Verification of Completeness

Recall that a RBS is considered to be *complete* if there exists at least one rule succeeding for any possible input state specification. Hence, a complete system is one able to react for any input.

In literature [3, 103, 123] there are two basic approaches to completeness verification. The most popular one is based on exhaustive enumeration of possible input data and systematic inspection of a given set of rules versus a table containing all possible parameters and conditions combinations. This kind of approach can be called an *exhaustive completeness check* [3]. Some examples of this approach are presented in [103, 123].

The other approach is based on a run-time validation of the expert system with the use of selected set of test cases [124]. Selected test problems should also provide an exhaustive list of possible cases. Some other approaches of this kind are also discussed in [3].

In [53, 54, 56, 57] a more general, first-order logic based approach is presented. The approach does not require exhaustive enumeration and testing of possible cases; instead, a proof-like procedure based on backward dual resolution is put forward.

In the following subsection logical (total) completeness, specific (physical) completeness and detection of missing rule preconditions will be discussed in turn. In all subsections the same set of rules will be considered, i.e.:

$$\begin{array}{lll} r_1: & \phi_1 & \longrightarrow h_1, \\ r_2: & \phi_2 & \longrightarrow h_2, \\ \vdots & & \\ r_m: & \phi_m & \longrightarrow h_m, \end{array}$$

or equivalently, given by (8.2) on page 138. Further, note that in fact only preconditions of the rules are of interest for completeness verification.

16.2.1 Logical Completeness of Rule-Based Systems

The approach proposed here comes from purely logical analysis [54, 60, 67] and does not require exhaustive enumeration of possible cases; instead a proof-like, algebraic procedure is put forward.

Consider the joint disjunctive formula of rule precondition of the form

$$\Phi = \phi_1 \vee \phi_2 \vee \ldots \vee \phi_m.$$

The condition of logical completeness for the above system is:

$$\models \Phi, \tag{16.1}$$

which simply means that Φ is a tautology. In first order logic based systems the proof can be performed directly with the use of backward dual resolution [53, 54]. However, recall that in tabular RBS the negation is usually not present explicitly. This means that no tautology of the type $\alpha \vee \neg\alpha$ can be present, and thus reduction to such type of tautology is not possible. Instead, reduction taking into account limited domains of system attributes can take place. However, contrary to the reduction operation aimed at minimizing the number of rules (and thus applied only to rules having identical conclusions), this time reduction can be applied to all the rule preconditions, disregarding their conclusions. Thus, the purely logical condition (16.1) can be replaced by a practical condition of the form

$$\mathit{Red}(\mathbf{\Phi}) = \square \tag{16.2}$$

where $\mathit{Red}(\mathbf{\Phi})$ is the maximal reduction of table $\mathbf{\Phi}$ and $\square$ denotes an empty table.

16.2.2 Specific Completeness of Rule-Based Systems

In most of practical cases the analyzed system may be logically incomplete. It can be designed to work in a certain limited context Ψ. The restrictions may follow from physical limitations on system parameters, specific 'local' character of the situation to be served, a limited knowledge of the system designer or may be the physical consequence of infeasibility of certain inputs.

Let Ψ denote the operating context for the above system. The *specific* (partial) *completeness condition* can be stated as follows

$$\Psi \models \Phi \tag{16.3}$$

where Φ denotes the disjunction of precondition formulae. Again, from logical point of view the verification could be purely logical and it could be proved by means of automated deduction (e.g. by direct use of bd-resolution [53, 54]). However, taking into account the tabular knowledge representation, an algebraic method for specific completeness verification would be suggested.

First, instead of considering m rule preconditions, one can apply maximal reduction of the formulae; most likely partial reduction will be applied, and the operation will result with a smaller set of k rules. Formally, we have $Red(\mathbf{\Phi}) = \mathbf{\Phi}^*$, where $\mathbf{\Phi}^*$ is the reduced k-rows table. As above, reduction is carried out disregarding the rule conclusions (which are different from one another), i.e. *any* two (or more) precondition formulae can be selected for reduction.

Now, three outputs are possible: the resulting table can be empty (no problem, full logical completeness holds), it can have exactly one row, or it can have k rows, where $k > 1$.

The case when $k = 1$ is again simple: the completeness check expressed by (16.3) can be replaced by

$$\Psi \models \Phi^*. \tag{16.4}$$

Since Φ^* is a single-row formula ($k = 1$) the check of (16.4) is equivalent to subsumption checking; thus it can be performed with the use of Subsect. 13.2.2. If C consists of a several row table (logically: a disjunction of several simple formulae), then the subsumption check must hold for any row of Ψ.

The most complex is the case of $\mathbf{\Phi}^*$ being a table of more than one row (and perhaps further irreducible). In such a case, the check should be performed for any row separately, and all the rows of Ψ must be covered. Let Ψ be the disjunction of the form $\Psi = \psi_1 \vee \psi_2 \vee \ldots \vee \psi_c$, where ψ_i denotes the *i-th* row of Ψ, $i = 1, 2, \ldots, c$. The sufficient condition for partial completeness takes the form

$$\forall i \in \{1, 2, \ldots, c\} \quad \exists j \in \{1, 2, \ldots, k\}: \quad \psi_i \models \phi_j^* \tag{16.5}$$

where ϕ_j^* denotes the $j - th$ row of table $\mathbf{\Phi}^*$. The geometric interpretation of condition (16.5) is straightforward — any elementary context ψ_i must be covered by precondition formula of a rule.

Note that during reduction of the formulae also more specific formulae can be used if necessary (i.e. subsumed ones). This may be crucial if some two formulae taken together cover certain formula ψ_i, but they are irreducible in a direct way (e.g. some of their conditions are different; in such a case specialization of some conditions may be necessary to obtain identical parts of the formulae).

An alternative approach may be based on splitting the context defining formula Ψ into smaller parts, and this seems to be more straightforward solution from computational point of view. In such a case, the separation values of selected attributes should be carefully selected so as to avoid too detailed split. A reasonable solution may be accomplished by selecting the values occurring in $\mathbf{\Phi}^*$, i.e. ones which still exist in the description language after maximal possible reduction. This means that characteristic values of attributes can guide the split operation.

In practice, the splitting operation of every simple formula ψ_i (uncovered by at least one ϕ_j^*) can be performed by the procedure outlined below:

- for any attribute A consider all its values (sets or intervals) appearing both in ψ_i and the table $\mathbf{\Phi}^*$; denote the values as $V_1, V_2, \ldots, V_m$;
- generate partition of the domain of A as a collection of sets $\mathbf{B} = \{B_1, B_2, \ldots, B_n\}$ (the so-called *blocks*), such that $B_i \cap B_j = \emptyset$ for $i \neq j$ and $B_1 \cup B_2 \cup \ldots \cup B_n = D_i$, where $i, j \in \{1, 2, \ldots, n\}$, the biggest ones but such that for every V_i, $i = 1, 2, \ldots, m$ there exist $B^1, B^2, \ldots, B^k \in \mathbf{B}$ such that $V_i = B^1 \cup B^2 \cup \ldots \cup B^k$, i.e. every value V_i can be expressed as a sum of a certain selection of blocks of $\mathbf{B}$; generation of $\mathbf{B}$ for intervals can be done by finding all intersections of the type $V_i \cap V_j$, $i, j \in \{1, 2, \ldots, m\}$, but for nominal sets also intersections of three, four, five, etc. sets should be considered as possible candidates (any superset is immediately eliminated);
- split every value of selected attribute A_j in ψ_i into largest possible sets being sums of blocks, say $B^1, B^2, \ldots, B^{j_i}$ and replace a particular row in ψ_i with appropriate j_i rows;
- repeat the procedure for every attribute A in ψ_i;
- for the resulting table being the output of the procedure apply the check following from condition (16.5);
- the system is specifically complete if every row is covered by a row of Φ.

To illustrate the idea one can employ a simple graphical interpretation for visualization of the completeness check. Every rule — being a row of the table — forms a path covering a certain selection of attribute values in the *n-th* dimensional attributes space (see Fig. 16.1). The values of specific attributes generate partitions of the domains of attributes.

Further, note that after performing the reduction of rule preconditions (if possible, a maximal one), the rules tend to cover bigger areas of selected attributes. In some best cases, certain attributes may also turn out to be unimportant, since all the values of them are covered.

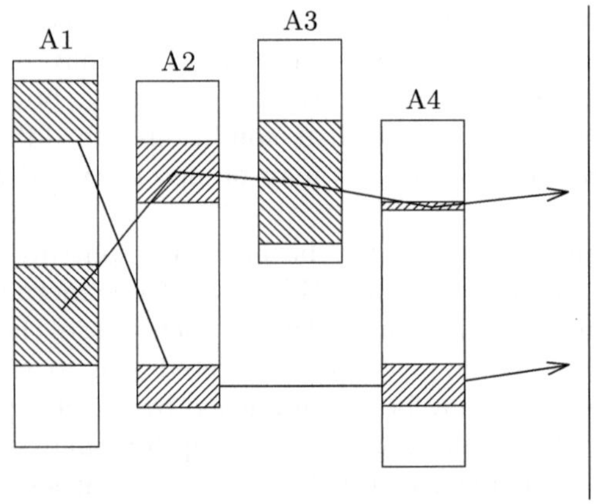

Fig. 16.1. The concept of attribute space and interpretation of a rule as a path covering selection of values

Note that in case of finite domains, the split could be done by taking into account single elements of the attribute domains; this, however, would be equivalent to exhaustive enumeration of the input cases. The power of the algebraic approach lies in operating on sets of input cases which can be achieved by a high-level split (not too detailed one).

Finally, let us assume that the values of attributes to be covered were split w.r.t the partition, e.g. given by boundary values induced by the values appearing in rule preconditions and the constraints, after the rules have been maximally reduced (in fact, the preconditions of rules were reduced to a minimal number of maximally general formulae). Now, the check for completeness consists of tracing if all the combinations resulting from the split of the context constraint are covered by some of the 'thickened' paths generated by the reduced set of rules.

The graphical interpretation is on the Fig. 16.2.

In general, the check may suffer from combinatorial explosion; however, due to operating on blocks (sets, intervals) rather than on individual values, the number of detailed checks is significantly reduced with respect to the methods based on exhaustive enumeration.

16.2.3 Missing Precondition Identification

Determination of missing preconditions and thus missing rules can be based on the above scheme for completeness verification. As before, assume that Ψ is a tabular form of a formula specifying the required area to be covered with preconditions of the rules. Further, let $Red(\mathbf{\Phi}) = \mathbf{\Phi}^*$ denote the maximally reduced table of formulae preconditions. Considering any of the attributes, say A_i, let V_i denote the set of characteristic values of this attribute, still occurring in the preconditions after reduction. These may be boundary values defining

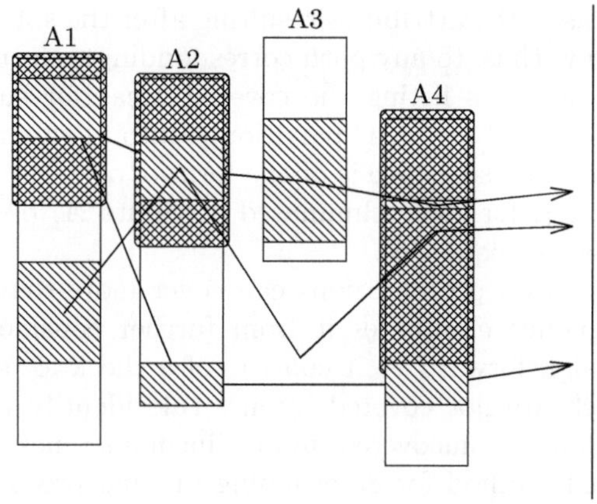

Fig. 16.2. The idea of covering the attribute space with more general, reduced set of rules

boundaries of intervals (in case of ordered attributes) or simple subsets of attribute domain (in case of attributes for which their domain is an unordered set). Note that the stronger reduction is possible, the less characteristic values are still left.

Now let us use the values V_i to split the domain of any attribute into (possibly maximal) intervals or subsets. In case two intervals or subsets overlap, a third interval or subset can be distinguished, so that the split forms a partition — no two intervals/subsets overlap, and their sum gives the complete domain.

A simplified graphical interpretation of the idea is presented in Fig. 16.3.

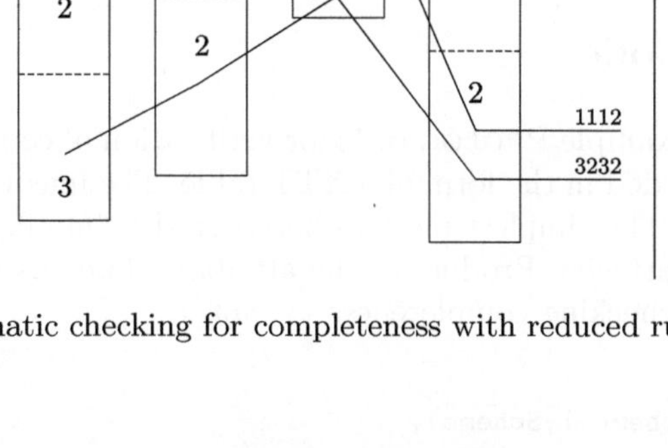

Fig. 16.3. Systematic checking for completeness with reduced rules in the space of attributes

The subareas of the attributes resulting after the split can be numbered in an arbitrary way; thus, to any path corresponding to a rule there corresponds a sequence of numbers listing the covered areas for subsequent attributes. If the domain of attribute A_i is partitioned into k_i subareas (and there are n attributes), then any sequence is of the form $\alpha^1\alpha^2\ldots\alpha^n$ (of length n), where $\alpha^i \in \{1,2,\ldots,k_i\}$; for some eliminated attribute A_j $\alpha^j = \_$. There are also $k_1k_2\ldots k_n$ possible sequences.

Note that certain preconditions can cover more than one such sequence; covering a sequence eliminates it from further consideration. The test for covering is straightforward — it consists of a check for subsumption. All the sequences which are not covered by any rule identify potential gaps in the system, i.e. define the uncovered inputs (incompleteness).

The outlined method for determining missing preconditions (possible for further analysis) is constructive, but in case of larger systems may suffer from combinatorial explosion. This effect, however, is significantly minimized here in comparison to the approaches based on direct exhaustive enumeration. This is so because of the following reasons.

Firstly, maximally reduced form of precondition table $\mathbf{\Phi}^*$ is used only; this restricts the division of any attribute to limited number of areas, and not to all its possible values.

Secondly, the method is aimed at analyzing a local system, operating in some well-defined, rather narrow context Ψ. This is possible thanks to introduction of hierarchical structure of the system.

Thirdly, splitting Ψ is necessary only w.r.t. its rows which are not initially covered by $\mathbf{\Phi}^*$. Finally, for reduced preconditions where the values of certain attributes are unimportant (the attributes are eliminated), the number of potential paths covered is multiplied by the factor equivalent to the number of areas to which the attribute was split.

Note also, that from logical point of view, the way of determining preconditions for missing rules may be considered to be equivalent to generating the $\neg(\mathbf{\Phi}^*)$ formula.

16.3 Verification of Completeness in XTT — a Prolog Code

Consider an example PROLOG code for verification of completeness of attributive rules encoded in the form of a XTT table. The uncovered states are listed as an output. The simplest method based on determining all the input states as result of Cartesian Product of the attribute domains is applied. The main predicate for checking completeness is *cmp/1*.

```
cmp(T):-
        scheme(T,Scheme),
```

```
        genstate(Scheme,State),
        covered(T,State),
        fail.
cmp(T):-
        write('No more uncovered states by  table'),
        write(T), nl.

genstate(Scheme,State):-
        Scheme = [A],!,domain(A,D),
        member(V,D),State=[f(A,atomic,V)].
genstate(Scheme,State):-
        Scheme = [A|Atts],
        genstate(Atts,StateAtts),domain(A,D),
        member(V,D),State=[f(A,atomic,V)|StateAtts].
covered(T,State):-
        assertlist(State),
        rule(T,_,P,_,_,_,_,_),
        satisfied(P),
        retractall(f(_,_,_)),!.
covered(T,State):-
        write('Uncovered state: '),write(State),
        write('by table'), write(T),nl.
```

The possible input states are generated in turn to avoid memory overflow; they are also sequentially checked for covering by at least one rule. The covering is translated to precondition satisfaction for at least one rule.

Predicate *scheme/2* reads the scheme definition for table T (the sequence of attributes used in preconditions). Then the predicate *genstate/2* generates all possible states and asserts them into the global memory. Finally, predicate *covered/2* checks if a given state is covered by a certain rule; if not, the state is reported as an uncovered one.

The auxiliary predicates are specified below.

```
cmp(T):-
assertlist([]):- !.
assertlist([F|R]):- assert(F), assertlist(R).

satisfied([]) :- !.
satisfied([Fact|Facts]) :-
        valid(Fact),
        satisfied(Facts).

valid(f(A,atomic,V)) :- f(A,atomic,V),!.
valid(f(A,natomic,V)) :- f(A,atomic,W), V \== W, !.
valid(f(A,set,Set)) :-
        f(A,atomic,V),
        set(Set,SetValue),
        member(V,SetValue),!.
valid(f(A,set,Set)) :-
```

```
          f(A,set,SA),
          set(SA,SAValue),
          set(Set,SetValue),
          subset(SAValue,SetValue),!.
valid(f(A,nset,Set)) :-
          f(A,atomic,V),
          set(Set,SetValue),
          \+ member(V,SetValue),!.
valid(f(A,nset,Set)) :-
          f(A,set,SA),
          set(SA,SAValue),
          set(Set,SetValue),
          intersection(SAValue,SetValue,[]),!.
valid(f(A,interval,i(B,E))) :-
          f(A,atomic,V),
          V >= B,
          V =< E,!.
valid(f(A,ninterval,i(B,_))) :-
          f(A,atomic,V),
          V < B,!.
valid(f(A,ninterval,i(_,E))) :-
          f(A,atomic,V),
          V > E,!.
valid(f(A,interval,i(B,E))) :-
          f(A,interval,i(BB,EE)),
          BB >= B,
          EE =< E,!.
valid(f(A,ninterval,i(_,E))) :-
          f(A,interval,i(BB,_)),
          E < BB,!.
valid(f(A,ninterval,i(B,_))) :-
          f(A,interval,i(_,EE)),
          EE < B,!.
```

A more advanced approach to verification of completeness may incorporate the ideas of *granular sets* and *granular relations* [72, 73, 74]; instead of single states blocks of states are analyzed at a time. In the above code one could easily modify the definition of domains of attributes so that granular values will be taken into account.

Part IV

Design of Rule-Based Systems

17

An Introduction to Design of Rule-Based Systems

This chapter presents basic methodological issues concerning the design of rule-based systems[1]. The most common problems encountered during practical design and implementation are outlined. An outline of structural approach to the design process is presented in brief.

17.1 Problems of Rule-Based Systems Design

Rule-based systems are used extensively in practical applications, especially in domains such as automatic control, system monitoring, technical and medical diagnosis, etc. [48, 129].

Many modern applications in various domains are reported in [51]. The rule-base technology is useful and efficient in numerous areas requiring symbolic processing of knowledge. Its success is mostly due to extremely powerful but simultaneously straightforward and transparent operational knowledge specification.

However, although the rule-based programming paradigm seems relatively conceptually simple, in case of realistic systems it is a hard and tedious task to design and implement a rule-based system that works in a correct way. Problems occurs as the number of rules exceeds even relatively very low quantities. It is hard to keep the rules consistent, to cover all operation variants and to make the system work according to desired algorithm.

Particular problems concern the selection of knowledge representation formalism as well as design of an appropriate rule-base and knowledge acquisition. Further problems include building the inference engine and developing control strategy.

[1] I gratefully acknowledge that the analysis presented in this chapter was worked out by Grzegorz J. Nalepa and his findings are presented in his Ph.D. thesis [92]. Moreover, I gratefully acknowledge that Sects. 17.2, 17.3, and 17.4 are largely based on excerpts from his Ph.D. thesis [92].

A. Ligęza: *Logical Foundations for Rule-Based Systems*, Studies in Computational Intelligence (SCI) 11, 231–240 (2006)
www.springerlink.com

Last but not least there are problems with verification, validation and testing of the knowledge-based systems [135]. Building a non-trivial system[2] requires solving several methodological issues and the use of specific software tools [51].

The main problems encountered in the development of knowledge-based systems in general and expert systems or rule-based systems in particular are located in the following areas:

- *knowledge representation*, which concerns selection of a proper language for encoding the acquired knowledge;
- *knowledge acquisition*, which is the process of extracting domain knowledge possessed by human expert;
- *developing inference mechanism*, which concerns design and implementation of an interpreter capable of rule execution;
- *developing reasoning control*, which concerns a meta-level knowledge for organizing the search and order of rules during reasoning so as to perform in an efficient way and avoid exponential explosion, dead-ends or infinite loops, etc.;
- *knowledge verification*, which consists of checking certain characteristics and correctness of the system knowledge base;
- *explaining solutions*, which concerns human/computer interaction, presenting solutions to the user, as well as explanations *why* and *how* the solutions have been found;
- *developing man-machine interfaces*, which concerns human/computer interaction during design, knowledge acquisition, inference and verification.

In numerous knowledge-based systems the basic, core knowledge representation is selected to be the one with rules. By using a rule-based system specified in an appropriate logic-based language the knowledge base is defined in a declarative manner. The basic language is normally one being equivalent to attribute logic or some more elaborated versions of it. With respect to inference mechanism usually forward chaining is used in monitoring and control systems, and backward chaining in diagnostic and some decision support systems. Examples of some well-known solutions, both with respect to basic theory and practical examples of such systems, are presented in the domain literature [44, 46, 48, 51, 84, 129].

A practical *implementation* of rule-based systems encounters two main problems. The first one is known as the *knowledge acquisition bottleneck* and it concerns the well-known difficulties with obtaining a precise knowledge specification. Using specific *knowledge representation structures* and *abstract*

[2] Some sources say that a non-trivial system is one having several hundreds of rules; some others say that even fifty rules may be difficult to handle. Without referring to the exact number of rules one may say that a non-trivial rule-based system is one outperforming average human in a specific domain. Disregarding the number of rules, this simultaneously means that the *knowledge* encoded in the system is non-trivial.

knowledge representation may be used to help to overcome this problem (see Chap. 18). This is so thanks to systematic guiding the development of the rule-base by introducing structure for ordering the design.

The second problem concerns the analysis, verification and validation of knowledge. By making the analysis of the knowledge base a part of ongoing knowledge acquisition and review process, expert system developers can minimize the time and resources devoted to development of such systems. In order to assure safe, reliable and efficient performance, analysis and verification of selected qualitative properties should be carried out [3, 6, 20, 21, 22, 81, 101, 109, 110, 111, 134, 135, 136]. Those properties include features such as, completeness, consistency and determinism. However, verification of them *after the design* of a rule-based system is both costly and late. The verification may be complex, so in most of practical applications building, debugging and maintaining the rule-base are the most costly activities. Below, a short analysis of these problems from the *Knowledge Engineering* point of view is carried out.

17.2 Knowledge Engineering

Knowledge is a key issue in development of modern information systems. The process of extracting and encoding domain knowledge held by human experts is called *knowledge engineering* [50]. Given the state–of–the–science in AI today, knowledge engineering remains a time–consuming and labor–intensive process wherein an AI technologist, called a *knowledge engineer*, must repeatedly interview one or more human experts over a long time period to extract the heuristics to be encoded in the expert system knowledge base.

Knowledge engineering is essentially a process of acquisition and abstraction on one hand, and design and construction on the other. The knowledge engineer must not only learn how the expert solves problems, but must learn it to such a level of detail that he or she can in turn provide precise instructions for a *tabula rasa* machine. This is not an easy task since human experts are often unaware of their own internal reasoning process and possessed skills. The top-level expertise remains often unconscious and informalizable, especially with respect to the origin of provided solutions.

That is why, when building and using expert systems, a number of important people are involved:

- the *human expert*,
- the *knowledge engineer*,
- the *system developer* [89].

There are many factors that can make the knowledge engineering process difficult [50]:

- miscommunication between human expert and knowledge engineer,
- a significant start–up time needed for the engineer to achieve a certain level of experience with the domain,

- limitations on availability of human experts,
- choosing best knowledge representation for the domain.

Knowledge base created during knowledge engineering process may prove very useful for a number of other applications, than building an expert system:

- the knowledge base can be considered a model for validation of forecasts,
- the knowledge base can be seen as institutionalized archives of the domain,
- the knowledge base may also be a good training facility for other people (i.e. new human experts).

17.2.1 Knowledge Acquisition

An appropriate knowledge representation formalism can be selected on the base of analysis of the nature of knowledge to be encoded as well as particular excerpts of knowledge provided by human experts. Once the appropriate knowledge representation scheme is selected, the process of knowledge acquisition should be performed.

According to [14] *knowledge acquisition* is the transfer and transformation of potential problem-solving expertise from a knowledge source to a program. The objective of *knowledge acquisition* is to conceptualize what the expert knows and how he or she solves problems.

A brief categorization of some kinds of knowledge includes:

- *declarative knowledge*, i.e. static statement of facts and rules, referred to as *'know that'* (facts or heuristics are good examples of this knowledge);
- *procedural knowledge*, i.e. operational knowledge (algorithms), refers to *'know how'* (involves an automatic response to a stimulus);
- *semantic knowledge* which refers to the system under analysis, its properties and behavior; it can be described as deep knowledge an expert has, often based on knowledge of *model* of the system; it is composed of different facts, definitions, and relationships among them;
- *episodic knowledge*, refers to recorded observations and experimental information;
- *meta-knowledge* which allows an expert to choose a solution for the problem and evaluate its reliability, it also lets an expert detect an unsolvable problem [116].

There are three distinct approaches to acquiring the relevant knowledge from a particular domain:

1) The knowledge is extracted from a domain expert.
2) The builder of the knowledge base is a domain expert.
3) The system learns automatically from examples [44].

The first approach is commonly used, however it poses some difficulties. There are several acquisition techniques (see [116]), such as:

- *interview* where an expert acts as a lecturer and a knowledge engineer asks questions to clarify understanding of the problem;
- *domain and task analysis*, an engineer obtains knowledge from direct observations of an expert performing his or her task;
- *simulations*, where an expert demonstrates his or her expertise in a computer–simulated environment, which is fully controlled by an engineer;
- *automated tools*, where an expert may automate knowledge acquisition process with certain computer–based tools, such as CASE tools.

These techniques involve many different skills of a knowledge engineer. They may also require not just engineering but also a certain psychological practice. The main way of extracting the knowledge from an expert is by interview. This is prone to communication difficulties.

In the second approach a knowledge engineer becomes a domain expert, or a domain expert becomes a knowledge engineer. The third approach takes place when the system is able to learn and generate its own knowledge. This approach is often used when the knowledge is unknown or difficult to express explicitly.

17.2.2 Knowledge Verification

In order to assure safe, reliable and efficient performance, the analysis and verification of selected qualitative properties should be carried out [3, 20, 21, 22, 134, 135]. Those properties include features such as, completeness, consistency and determinism. The verification problems concerning rule-based systems are discussed in detail in Chap. 12 and some following sections.

17.2.3 Knowledge Management

Knowledge management is a new term coined in the business organizations to reflect some old concepts, such as sharing knowledge in a collaborative way to stimulate new ideas, combined with modern computer-based intelligent tools and methods [52]. It is strongly rooted in knowledge engineering, and these two fields often overlap. Knowledge management however emphasizes the aspect of sharing, applying, extending and maintaining knowledge obtained in the engineering process.

17.3 Design of Rule-Based Systems: Abstract Methodology

Practical design of non-trivial rule-based systems requires a systematic, structured and consistent approach. Such an approach is usually referred to as a *design methodology*. The basic elements distinguishing one methodology from the other are the *internal design process structure* i.e. the way of structuring the design process and the *components* of the process. The structure can be linear, linear with loops, hierarchical (top-down, bottom-up), etc. The components may be various procedures, techniques, tools and documentation aids to support and facilitate the process of design [23].

The *methodological pyramid* (see Fig. 17.1) introduced by Wielinga et al. [140] is a convenient way to present what is involved in a methodology.

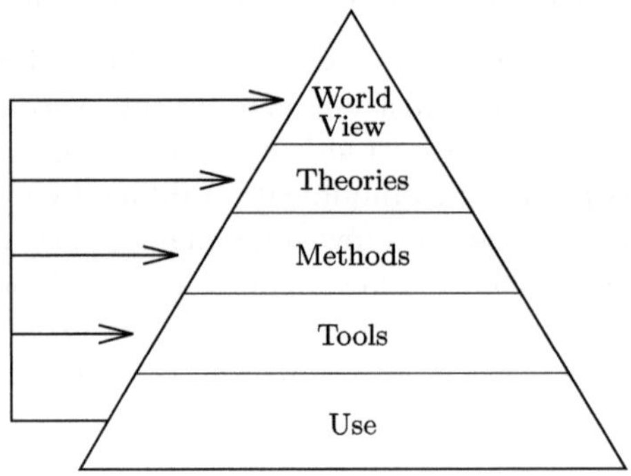

Fig. 17.1. The methodological pyramid

Based on the idea of methodological pyramid and the scope of the methodology, Liebowitz [51] places methodological approaches to expert systems design into three categories:

1) *life-cycle*,
2) *focused*,
3) *full-fledged.*

A classical *life-cycle* approach to the expert systems development process presented in Fig. 17.2 is based on general software engineering concepts (the so-called *Cascade* or *Waterfall model*). It was presented by Buchanan [14] and often cited, e.g. in [46, 51] (an explicit *verification* at the formalization stage is sometimes suggested [5]).

It consists of *five* stages (Fig. 17.2):

1. *Identification* includes identification of the class of problems the system is expected to solve, including the data the system will work on, and resources that are available.
2. *Conceptualization* consists of figuring the key concepts and relationships between them, such as kinds of data, flow of information, and underlying system structure.
3. *Formalization* involves the process of understanding, describing and formalizing the problem search space, and the way solutions are found; it should include the verification of information about the system.

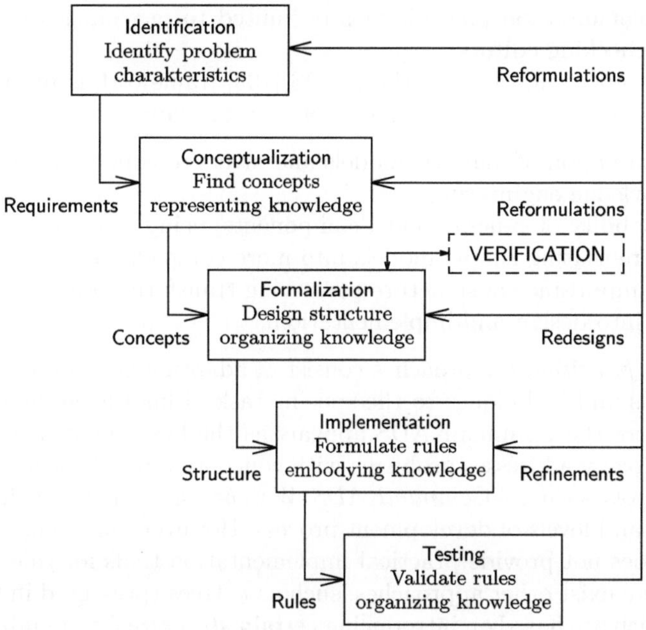

Fig. 17.2. Expert system life cycle

4. *Implementation* aims at turning a formalization of knowledge into a runnable program, including encoding rules and inference mechanism in some kind of a lower level programming language.
5. *Testing* tries to evaluate the result of implementation and possibly detect errors that were made during the design.

The *focused* methodologies are restricted to one or few stages of the design process and are oriented towards in-depth support of specific design activities included there. The main focus is typically on the knowledge acquisition phase which constitutes a well-recognized bottleneck.

There are three main directions of approaching the problem:

1) attempting to build a formal frame (a *model* for the expert knowledge using tools and techniques such as case analysis, interviews, protocols and various forms, etc.);
2) direct transferring of the expert knowledge to the knowledge base;
3) constructing a domain-specific environment applicable in some narrow domain.

Several *focused* methodologies for building knowledge-based systems have been studied and proposed as conceptual tools for simplifying the design of such systems [27], with KADS [139] as a standard example. Such methodologies support mainly subsequent stages of the *conceptual design* in case of large systems, while direct technical support of the logical design and during the implementation phase is mostly limited to providing a context-sensitive syntax checking editors.

The motivation behind the KADS [139] framework is primarily the management of complexity. It is based on the following principles [46]:

- introduction of multiple models and layers to cope with the complexity of knowledge engineering;
- reusability of generic model components, using top-down approach;
- differentiating simple models into more complex ones;
- the importance of structure-preserving transformation of models of expertise into design and implementation.

The *full-fledged* approaches consist of adapting general software engineering tools and techniques to the specific task of knowledge-based systems design. Here the CommonKADS appears as the best known example [27]. The methodology addresses problems with conceptual and logical analysis. It provides tools such as *CommonKADS Workbench* supporting different system models and levels of development process. However, due to its general character it does not provide practical implementation tools for rule-based system.

There exist other approaches, such as ψ-trees (presented in the next chapter) which are based on introducing certain *structure* for organizing the knowledge and guiding the design process. They put emphasis on proper *formalization*, and suitable *representation*. They are especially useful in designing technical systems in well-defined domains, such as control rule-based systems, diagnostic or monitoring ones.

ψ-trees is a knowledge representation method created with a design approach in mind. It was discussed in [58, 59]. It is based on a complex variant of decision trees. It is focused on structuring the knowledge into modules, each operating in the context defined by ψ. In ψ-trees a single node may have more than two links which makes the decision process more realistic, and allows to compare attributes with many different values at a single node. The structure of such trees is modular and hierarchical. An attempt to build a CASE tool supporting ψ-trees was presented in [69]. The idea is novel, so it lacks effective tools for practical implementation of rule-based system. The ψ-trees based approach will be presented in Chap. 18.

17.4 Rule-Based Systems Design: Basic Stages

As it was mentioned before, in order to assure safe, reliable and efficient performance, analysis and verification of selected qualitative properties should be carried out [3, 6, 20, 21, 22, 81, 101, 109, 110, 111, 134, 135, 136]. These properties include features such as, completeness, consistency and determinism. However, verifying them *after the design* of a rule-based system is both costly and late. Unfortunately, methodologies such as the ones mentioned above do not support this stage in extent.

Certain sources, such as [46], give only certain practical advice for avoiding common problems in rule-based system development, such as:

- organize and group rules into rule sets,
- use incremental development process with testing and validating on early stages,
- design for transparency from the beginning of the development process.

In some cases *database design approach*, presented in [23], could be adapted for rule-based systems design. It is a top-down structured approach for analyzing and modeling a set of requirements for a database in a standardized and organized manner.

In its standard form it consists of *three* main phases:

1. *Conceptual design.* This stage is focused on creating a conceptual model of the system to be built, it is developed with strong reference to requirements specification; the model is validated against these requirements; it is the source of information for the logical design phase. Conceptual design stage includes identification of important entities and relations among them, their characteristics, etc. but typically does not define specific knowledge representation or inference control mechanism.
2. *Logical design.* This stage concerns translation of the conceptual representation to the logical structure; results in the creation of the logical data model of the problem. As the result — in case of rule-based systems — a set of rules, specified in a selected abstract language is obtained.
3. *Physical design.* This is the phase during which implementation decisions on how the logical structure will be implemented are made; the logical design must be further translated into the language of selected implementation tool. There is a certain level of feedback with logical design phase.

Considering the internal structure it is also similar to the well-known waterfall model of software development presented in Fig. 17.2; however, that model is too abstract to prove useful during the rule-base design process. Conceptual and logical design phases are iterative, with multiple refinements. To some extent they may be seen as a learning process.

System designers learn to better understand the system they are building. These phases are critical to the success of the system.

The design methodology proposed in this work uses some ideas taken from the above approach. In Chap. 18 logical foundations are presented and the concept of ψ-trees for efficient representation of the logical structures of the system under design is put forward. Then, in Chap. 19 the core approach to the design of XTT systems is presented. In the last Chap. 20 an example of the design is presented.

This Section is devoted to outlining the main stages for design of rule-based systems. Until now there is no unique, consistent and universal methodology for designing such systems.

However, it seems that the design process can be structured into interleaving stages and it must cover the following principal phases:

1. Specification of the desired *goal* and *functionality* of the system.
2. Selection of knowledge representation language (e.g. propositional logic, attributive logic, XTT, etc.).
3. Definition of an outline of the system structure and operation.
4. Definition of system components (e.g. XTT tables, rules) and relations/connections among them.
5. Specification of inference control mechanism.
6. Knowledge acquisition and encoding.
7. Verification of formal properties and optimization of the code.
8. Implementation and testing and evaluation of work of the system with regard to the specified *goal* and *functionality*.

Obviously, the above stages can interleave, and can be repeated, perhaps following the spiral model of software development. Further, and it is perhaps one of the most important issues, the design should be structured into a *hierarchical approach* — the design activities should be split into several levels of abstraction and at each level various degree of details is required.

Note, also, that the above activities belong to two different, separate groups; these are the *constructive activities* concerning definition and development of knowledge elements and inference mechanism and *verification activities* oriented towards assuring correct work of the system. Now, in general, these two stages can be organized in a classical way, i.e. one after another or through interleaving (and this is the proposed approach to support design of rule-based systems), so in the following two ways:

1) static, ex-post verification of the partially or completely developed system in order to suggest improvements; such a procedure can be repeated in several cycles;
2) on-line, dynamic support of design by instant detection of anomalies occurring during the design process and generation of appropriate information for the developer of the system.

The first approach is rather obvious and may be applied in a rather straightforward way; in fact, to certain degree it is a part of todays practice in development of rule based systems. However, it suffers from obvious disadvantages, including repetition of the design procedure. Moreover, errors in design discovered after completing an edition of the knowledge base may result in hard to deal with problems, and improvement of the rules may cause new errors to appear.

The second approach, proposed initially in [58] and [59] seems to be much more attractive and efficient. However, it requires development of a special software tool supporting the design.

In the following sections an outline of selected stages is presented.

18

Logical Foundations: the Ψ-Trees Based Approach

This chapter is devoted to a presentation of a logically-founded framework for design of rule-based systems. It is based on the use of the concept of *ψ-trees* — a tool for structuring the design process. It is aimed at on-line, dynamic support of design by instant detection of anomalies occurring during the design process and generation of appropriate information for the developer of the system.

The presented approach was proposed initially in [58] and [59]. It requires development of a special software tool supporting the design.

In the following sections an outline of selected stages is presented.

18.1 An Intuitive Introductory Example

For intuition, let us consider a possible design process of a simple temperature control system. This will be a rule based system capable of applying the two-valued and three-valued switch type algorithms (relay type) if stabilization of a standard temperature is required, or controlling the temperature towards it maximal or minimal value if required. All of this takes place only if the control of temperature is required.

In order to simplify systematic design we shall use a tree-like graphical representation for displaying combinations of various conditions. The idea is similar to standard binary *decision trees*; however, the proposed construction is a bit more general and allows for use of any finite number of branching conditions, description of the branching conditions with first order level languages, and hierarchical design. The idea of such a tree corresponding to a part of the design process is presented on Fig. 18.1.

The outline of the design process can proceed as follows. We start with the root node and select a certain predicate (or more complex set of formulae) to define the initial branching.

A reasonable idea is to select a formula of relatively high level, i.e. defining as general conditions as possible, and providing contexts for further

A. Ligêza: *Logical Foundations for Rule-Based Systems*, Studies in Computational Intelligence (SCI) 11, 241–249 (2006)
www.springerlink.com

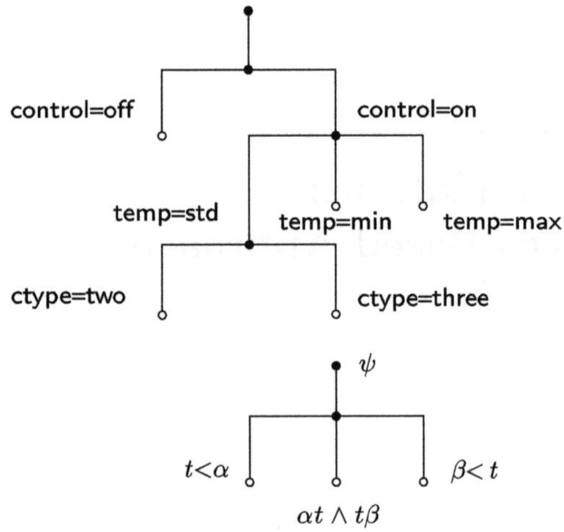

Fig. 18.1. An example ψ-tree for design

specification. In our case let it be the first condition specifying if the control is required (`control=on`) or not (`control = off`). Now, leaving the left branch temporarily pending, we further develop the right branch; note that now we are in the context defined by the formula `control = on`. The next selected formula may concern the mode of temperature regulation, i.e. if a steady temperature is required (`temp=std`) either the minimal (`temp=min`) or the maximal (`temp=max`) temperature should be achieved.

Now, continuing, leaving the two right branches temporarily pending (possibly for further continuation of design), we develop the left branch. Note, however, that now the context is defined by the conjunction of formulae assigned to the branches on the path from the root node to the current leaf node; in our case the formula is of the form `control=on` $\wedge$ `temp=std`. Now we have to select the controller type. Assume there are given two possibilities: a two-level one (`ctype=two`) and a three-level one (`ctype=three`).

One can proceed with the design in an analogous way, but a reasonable idea may be to split the tree into the mother initial tree and several subtrees whose roots are joined to the leaves of the mother tree. In Fig. 18.1 we leave the context defined by $\psi = [(\texttt{control} = \texttt{on}) \wedge (\texttt{temp} = \texttt{std}) \wedge (\texttt{ctype} = \texttt{three})]$ and start development of a new, smaller child-tree for the context defined by ψ. It will be referred to as a ψ-tree, and it describes the design process for the context ψ. In our case the branching condition refers to the current temperature, and we have three possibilities: $t < \alpha$, $(\alpha \leq t) \wedge (t \leq \beta)$, and $\beta < t$.

The final step is to assign to any leaf node a control action to be undertaken. For example, for the final result of the partial design process presented above (there are still pending leaf nodes to be developed), the following set of rules may be generated:

$$\begin{array}{ll} r1\colon \psi \wedge (t < \alpha) & \longrightarrow \texttt{set(heat,on)},\\ r2\colon \psi \wedge (\alpha \leq t) \wedge (t \leq \beta) & \longrightarrow \texttt{set(heat,off)},\texttt{set(cool,off)},\\ r3\colon \psi \wedge (\beta < t) & \longrightarrow \texttt{set(cool,on)}. \end{array}$$

Obviously, when operating in the context defined by ψ, the preconditions can be simplified to the formulae in parentheses.

Having in mind the above example it would be worthwhile to analyze some steps of the design process. Firstly, the formula defining the branching condition is selected to be possibly general and reasonable for the current design context; although there is no unique, well defined selection procedure, this seems to be a reasonable, expert-acquired way of design procedure. Secondly, defining the precise branching alternatives we take care to *cover* all the existing possibilities; from logical point of view, under the assumed technical interpretation I, the following conditions are satisfied:

$$\begin{array}{l} \models_I \texttt{control} = \texttt{off} \vee \texttt{control} = \texttt{on},\\ \models_I \texttt{temp} = \texttt{std} \vee \texttt{temp} = \texttt{min} \vee \texttt{temp} = \texttt{max},\\ \models_I \texttt{ctype} = \texttt{two} \vee \texttt{ctype} = \texttt{three},\\ \models_I (t < \alpha) \vee [(\alpha \leq t) \wedge (t \leq \beta)] \vee (\beta < t). \end{array}$$

In other words, we attempt at satisfying the local *completeness condition* when defining branching possibilities. Thirdly, we also try to define them so that the alternative formulae do not overlap, i.e. no two of them can be satisfied at the same time. In fact, from logical point of view we have:

$$\begin{array}{l} \not\models_I \texttt{control} = \texttt{off} \wedge \texttt{control} = \texttt{on},\\ \not\models_I \texttt{temp} = \texttt{std} \wedge \texttt{temp} = \texttt{min},\\ \not\models_I \texttt{temp} = \texttt{min} \wedge \texttt{temp} = \texttt{max},\\ \not\models_I \texttt{temp} = \texttt{std} \wedge \texttt{temp} = \texttt{max},\\ \not\models_I \texttt{ctype} = \texttt{two} \wedge \texttt{ctype} = \texttt{three},\\ \not\models_I (t < \alpha) \wedge (\alpha \leq t) \wedge (t \leq \beta),\\ \not\models_I (\alpha \leq t) \wedge (t \leq \beta) \wedge (\beta < t),\\ \not\models_I (t < \alpha) \wedge (\beta < t). \end{array}$$

Such procedure is justified since we would like to obtain a *complete* and *deterministic* set of rules. For intuition, as the precondition formulae are defined by conjunctions of formulae assigned to branches on paths from the root to the leaf nodes, at any node we cover all the reasonable branching possibilities; hence completeness is assured. Simultaneously, no two 'paths' can be validated as true at the same time, since they must be different at at least one node and due to exclusion of the branching conditions they cannot be true at the same time. These problems are referred to in certain detail throughout this book.

Finally, by splitting the design tree into a mother tree and several child-subtrees we achieve the possibility of performing hierarchical, two-level design. First, we design the meta-rules, allowing for the so-called *context-switching*. Executing such a rule defines a context (set of states) and a subset of all the rules such that only rules of the subset can be used for the currently selected context. Then, we define the sets of rules for any specific context by development of a ψ-tree for any context defined by ψ. Of course, this procedure can be repeated at several levels.

To summarize, the essence of the lessons learnt from the intuitive example can be summarized as follows:

- organize the design in a *hierarchical* way by defining different, separate but complementary *contexts* of work and perform the design of a component defining the work in any context separately,
- the basing underlying structure serving as a back-bone for the system can be a tree or a graph (acyclic one); in the above example it was proposed to use a specific tree, to be called *ψ-tree*,
- proceed in such a way that theoretical properties such as determinism and completeness are *preserved during design* rather than employ separate verification stage at the end.

These concepts will be further incorporated in the design of tabular rule-based systems with the XTT model. Below the logical concept of *ψ-trees* underlying the XTT model is outlined in some more detailed way.

18.2 The Ψ-Trees for Design Support

In this section basic concepts concerning a specific form of semantic trees (to be called ψ-trees) used for design of complete sets of rules for some context ψ are presented.

Let N denote a set of nodes. We assume the common definition of a tree (there is a distinguished *root node* with no ancestor and any other node n' belonging to the tree has exactly one ancestor). A tree will be denoted with t, and $t(N)$ will denote a tree built from nodes belonging to N; the set of all nodes of tree $t(N)$ will be denoted with $N(t)$. The root node of any tree t will be denoted with $root(\mathrm{t})$.

Now, let FOR denote a set of considered formulae, and let ψ denote a distinguished formula. By I we shall denote the intended interpretation.

Definition 91. *Let f denote any mapping of the form $f\colon N \longrightarrow FOR \cup \{\psi\}$. A ψ-tree is any finite tree $t(N)$ satisfying the following auxiliary conditions:*

- *$f(root(t(N))) = \psi$,*
- *for any $n \in N(t)$, $f(n) \in FOR$.*

For intuition, a ψ-tree represents different (some or all) possible branchings into different more detailed situations of the analyzed system for some stable context defined by ψ. Any node $n \in N(t)$ represents a situation described with conjunction of all the formulae assigned to the nodes belonging to the path from *root*(t) to n (including ψ and $f(n)$). Such formulae will be denoted with lower-case ψ or ϕ.

Roughly speaking, going down the tree along some path beginning in the root node we add some new requirements (defined by the formulae met on the way) to be satisfied in the described situation. Obviously, the greater the depth of a node is, the 'more particular' situation is described by the appropriate formula determined by the path from the root to this node. Finally, the paths ending with the leaf nodes determine a set of 'most detailed' situations within the context situation defined by ψ.

A ψ-tree can be used as a basic tool for guiding the generation of rules precondition by domain expert. The idea is to develop such a tree in a top-down mode in a systematic way. Alternative conditions referring to some predicate or formula are represented by branching in the tree. Along the paths from the root to leaves precondition formulae (conjunctions of the appropriate conditions) are synthesized.

Now, the most important idea lies in such a generation of the ψ-tree that the set of the most detailed situations described by all the formulae assigned to paths ending with leaf nodes 'cover' the initial situation defined by the context formula ψ. With respect to this problem the following definition is proposed.

Definition 92. *Let $\psi_1, \psi_2, \ldots, \psi_k$ denote all the conjunctive formulae assigned to all the paths from* root(t) *to the leaf nodes of some ψ-tree t. The tree is referred to as a* complete ψ-tree *iff* $\psi \models_I \psi_1 \vee \psi_2 \vee \ldots \vee \psi_k$.

Note that this definition assures in fact the specific completeness of the system (within the context defined by ψ). Further, the aim of the above definition is obvious — any object of the considered system belonging to the context defined by ψ will be eventually covered by the subsets of the universe (more precisely, by at least one of them) defined with formulae $\psi_1, \psi_2, \ldots, \psi_k$. This is stated with the following proposition justifying the use of complete ψ-trees for design of rule preconditions.

Proposition 1. *A complete ψ-tree assures specific completeness of the developed rule-based system with respect to ψ.*

With reference to the mentioned tool for proving completeness (i.e. bd-resolution), the goal of developing ψ-trees is obvious; a structure of a complete ψ-tree provides a straightforward strategy of bd-resolution theorem proof for the condition defining completeness, i.e. iff $\psi \models_I \psi_1 \vee \psi_2 \vee \ldots \vee \psi_k$. The following proposition provides a sufficient condition for completeness of any considered ψ-tree.

Proposition 2. *A ψ-tree t is complete if the following conditions hold:*

- *for any non-leaf node n being an immediate ancestor (parent) of some leaf nodes and a conjunctive formula ψ' determined by the path from root to this node there is*

$$\psi' \models_I \psi^1 \vee \psi^2 \vee \ldots \vee \psi^j, \tag{18.1}$$

i.e. the formulae satisfy the completeness condition (18.1) *and therefore it is possible to resolve* $\psi^1, \psi^2, \ldots, \psi^j$ *generating formula* ψ' *and,*
- *the generated reduced in such a way tree is still complete;*

here $\psi^1, \psi^2, \ldots, \psi^j$ *are the formulae assigned to all the child nodes of* n.

The proof is by induction with respect to the tree size. The second condition is necessary only if some of the variables occurring in ψ' occurs also in the formulae $\psi^1, \psi^2, \ldots, \psi^j$ (resolving the formulae may require substituting for a variable and thus, by influencing the path above, it may violate the completeness of the tree resulting from bd-resolution application).

For intuition, the construction of a complete ψ-tree provides a strategy for a proof (derivation) of the root context formula ψ from the set of all formulae determined by the paths from root to leaf nodes.

Further, let us take a closer look at the step of developing a tree t' from a tree t by extending one of the leaf nodes belonging to t with a set of its successors; in fact this is a basic step in the design process. The problem is that not necessarily all the formulae ψ^i determined by some paths in the extended tree must be satisfiable under the assumed set of interpretation I. Let $\psi \wedge \psi^1, \ldots, \psi \wedge \psi^i$ denote the satisfiable formulae, and let $\psi \wedge \psi^{i+1}, \ldots, \psi \wedge \psi^j$ denote the unsatisfiable ones (here the context defined by ψ is taken into account). Of course, there is no need to develop the initial tree t with respect to nodes $n^{i+1}, \ldots, n^j$ being the final nodes described by the unsatisfiable formulae $\psi \wedge \psi^{i+1}, \ldots, \psi \wedge \psi^j$; roughly speaking, pruning the unsatisfiable paths corresponds to deleting unsatisfiable formulae in a disjunction (the one of Def. 92). The 'partially' developed tree will be a complete tree as well; this is a weaker version of Proposition 2.

Proposition 3. *A* ψ*-tree* t *is complete if the following conditions hold:*

- *for any non-leaf node* n *(being an immediate ancestor of some leaf nodes) and a conjunctive formula* ψ' *determined by the path from root to this node there exist formulae* $\psi^1, \psi^2, \ldots, \psi^j$ *such that*

$$\psi' \models_I \psi^1 \vee \psi^2 \vee \ldots \vee \psi^j \tag{18.2}$$

  where $\psi^1, \ldots, \psi^i$ *are the formulae assigned to all the child nodes of* n *and such that* $\psi \wedge \psi^l$ *is satisfiable for* $l = 1, 2, \ldots, i$*, and where* $\psi \wedge \psi^l$ *are unsatisfiable formulae for any* $l = i+1, \ldots, j$*;*
- *resolving (hypothetical) of the formulae assigned to leaf nodes given by* (18.2) *leads to a reduced but still complete* ψ*-tree.*

The above proposition may contribute to significant reduction of both the length of precondition formulae and, with respect to size of the tree, the number of rules as well.

Further, the question of determinism should be raised. As follows from the presented analysis, a sufficient condition for determinism is constituted by unsatisfiability of the conjunction of any two precondition formulae. Note that if any branching in the tree incorporates not only complete but exclusive set of conditions, e.g. as in the introductory example (under the assumed interpretation), then no two precondition formulae can be satisfied at the same time. This can be recapitulated in the following proposition.

Proposition 4. *A ψ-tree t defines a deterministic set of rules if one of the following conditions hold:*

- *for any non-leaf node n,*
$$\psi^i \wedge \psi^j \tag{18.3}$$
is an unsatisfiable formula, i.e. no two formulae describing different paths can be satisfied at the same time (i.e. exclusive branching conditions are always specified; ψ^i and ψ^j are the formulae assigned to the child nodes of n), or
- *for any two paths going through a node with non-exclusive conditions immediately below, exclusive conditions (for the paths) are added at some node(s) below.*

Thus, the ψ-tree can be used for simultaneous generation of not only complete set of rules, but by considering exclusive conditions at any branching node deterministic set of rules is obtained at the same time. In case overlapping conditions have to be used at some branching (e.g. for conditions simplification), the specific paths may be marked during the design as ones potentially referring to nondeterministic rules. A further check can be done after the design.

In construction of deterministic and complete systems, an inevitable occurring problem is the one of combinatorial explosion; obviously, a complete and deterministic system with ground preconditions, and based on n logical conditions (e.g atomic formulae) should have 2^n rules. Note however, that in our case this problem is significantly reduced. Firstly, the representation language allows for representation of variables, so there is one rule for a subset of the states encoded with ground formulae. Secondly, some of the potential precondition formulae describe physically infeasible states, and thus they can be omitted; the completeness is still assured by Proposition 2. Thirdly, we discuss a theoretical approach and an ideal case; in a practical design one can agree to partial completeness and incomplete determinism. Further, the complexity problems can be significantly reduced by hierarchal design.

The problem of finding minimal representation (reduction) cannot be solved directly during generation of the tree in a general way. This is because only rules with the same conclusions are likely to be reduced; the conclusions, however, are assigned to the precondition formulae after the tree is generated. However, the use of ψ-trees for design allows for immediate *local minimization* of the number of generated rules in a straightforward way.

The main idea of minimization is based on joining two or more rules with 'slightly different' preconditions. The idea of *local minimization* refers to performing this operation at any branching node, directly after developing it — the current leaf nodes leading to the same rule action can be joined at once. Or, even better, we simply do not perform 'too detailed' branching, that is all! This seems to be reasonable and efficient enough in most cases, but in a general case a post-design minimization may be necessary.

Note that if any path in the tree is assigned a different conclusion or action, no reduction is possible. If two or more paths have the same conclusion, then they should be analyzed for possible reduction. Both reduction and partial reduction can be applied. Note that due to the completeness condition and validity of bd-resolution, reduction of rules does not violate completeness — the reduced set of rules must be complete as well. Further, if the set of rules was deterministic, the reduced set of rules will be deterministic as well.

18.2.1 OSIRIS — a Design Tool

Based on the ideas presented above, an experimental tool for development of rule-based systems named OSIRIS was designed and implemented [141]. This tool was also described in [68] and [70]. In [141] a new idea for graphical representation of combined tabular systems (and their parts) together with decision trees was developed. This is the so-called *tab-trees* or *tree-table* representation.

The main idea of the tool lies in building a hierarchy of tabular systems [71]. This hierarchy is based on the ψ-tree structure. Each row of a OAV table is right connected to the other OAV table. Such a connection implies logical AND relation in between.

The component tabular systems used in tree-table representation are divided into two kinds: attribute tables and action tables. Attribute tables are the attribute part of a classical OAT, Action tables are the action part. There is one logical limitation. While attribute tables may have as many rows as needed (the number of columns depends on the number of attributes), action tables may have only one row; it means that the specified action, or set of actions if there is more than one column, may have only one value set, which preserves consistency.

An example of a tree-table representation is given in Figure 18.2. Please note, that a tree-table representation is similar to Relational-Data-Base (RDB) data representation scheme, but it is a perfect example of a ψ-tree, as well.

The main features of the tree-table representation are:

- very simple, readable and engineers acceptable knowledge representation, remaining in part the RDB tables, composed with
- hierarchical, tree-like representation,
- highly efficient way of visualization with high data density,
- power of the decision table representation,
- analogies to the RDB data representation scheme.

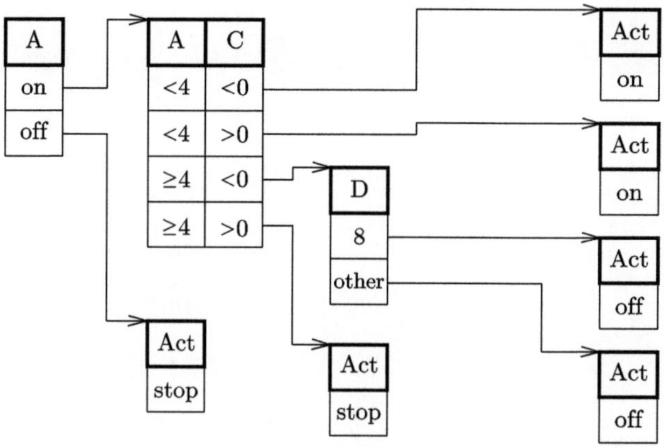

Fig. 18.2. An example of a tree-table representation

The OSIRIS tool is a multi-module system designed for UNIX environments (tested under Debian GNU/Linux, Sun Solaris 2.5). It consists of: a graphical environment for computer aided development of rules, a code generator for KHEOPS system, a validator, which provides completeness checking and a run-time environment for created rules. The architecture of the OSIRIS and further details are described in [141], and also in [68] and [70].

19

Design of Tabular Rule-Based Systems with XTT

In this chapter the main focus is paid to a new methodology for designing tabular rule-based systems with the use of the XTT knowledge representation paradigm[1]. From logical point of view, the structure of the design process follows the line of the ψ-trees; the design is hierarchical and structured into local components operating in well-specified contexts.

In order to facilitate the design we introduce the concept of *Attribute Relationship Diagrams* (ARD) for defining the functional dependencies among the attributes at an abstract level. The diagrams are used next to define the precise XTT structure of the system. The main innovation put forward in this chapter is introducing a combined ARD/XTT system as a design model for local components. In this way the global structure of XTT (Extended Tabular Trees) is generated. The material presented in this chapter is based on the recent developments presented in [92] and in [100]. The consequent stages of the design process are outlined in turn.

19.1 Principles the ARD/XTT Approach

The existing design methods and tools have limitations mostly in the following three areas:

1) knowledge representation methods,
2) framework for analysis and verification of system properties,
3) integrated computer tools supporting the design process.

To overcome these limitations a new approach based on knowledge representation with the eXTENDED TABULAR TREES (XTT) is proposed and

[1] I gratefully acknowledge that research for this chapter was undertaken by Grzegorz J. Nalepa and presented in his Ph.D. thesis [92]. Moreover, I gratefully acknowledge that Sects. 19.1, 19.2, 19.3, 19.4, 19.5 and 19.6 are largely based on excerpts from his Ph.D. thesis [92].

A. Ligęza: *Logical Foundations for Rule-Based Systems*, Studies in Computational Intelligence (SCI) 11, 251–261 (2006)
www.springerlink.com

developed. XTT were discussed in detail in Chap. 8, Sect. 8.5. In [92] a computerized tool called MIRELLA was presented and discussed. The architecture and implementation of this tool composed of an integrated visual editor based on the XTT approach, inference engine and verifier was also described there.

The ARD/XTT approach was invented with the goals to integrate the system development process from the *design* to *verification* stages; further, it supports the *implementation* through introducing the possibility of automatic code generation. Basing on it, an integrated rule-based system design and implementation process, supported by a computer tool, is offered in the following sections.

In Sect. 20 all the presented ideas concerning ARD and XTT are illustrated with an extended example.

19.2 Principles of the Integrated Design Process

The proposed approach follows the structural methodology for design of information systems. It is simultaneously a top-down approach, which allows for incorporating hierarchical design — in fact, any tabular component can be split into a network of more detailed components, and a network of components can be grouped together to form a more abstract component. The approach covers the stages of *conceptual*, *logical* and *physical* design. The principles of the integrated design process are based on selected existing approaches to system design, presented in Chap. 17.

The approach proposed herein does not aim at covering the whole *life-cycle*, as for example the one that can be found in [51], previously presented on Fig. 17.2. However, it does aim at including all phases of the system life-cycle from the *design* to *implementation* phase.

The following *three design phases* are identified:

1. *Conceptual design*, in which the basic structure of the system is identified, along with data and control flow, as well as main operating contexts, objects and their attributes; this allows for further defining the headers of XTT tables;
2. *Logical design*, which involves building table rows (corresponding to rules) and connecting tables; the XTT structure can be incrementally built, analyzed, and possibly verified and even optimized on-line at this stage;
3. *Physical design*, in which a preliminary *implementation* is done by building a PROLOG code (or any other target language e. g. the one of KHEOPS [42], since the approach is of generic character), which can be executed, compiled, debugged and possibly translated to system-specific representation.

This is a *top-down* approach. The names of design phases are similar to Relational Data Base design phases [23], introduced in Sect. 17.1 on page 231. However, the actual stages in each phase are different.

The issues that *are not* addressed by this process are:

- inference engine building — the process is mainly aimed at designing and testing the *rule base*, although a PROLOG-based inference engine can easily be developed (see Sect. 11.9); an example engine is provided in [92]; it can, however, be replaced by another implementation, if needed;
- user interface building — a simple shell is provided with the inference engine; it is sufficient for the current testing of the implementation; however, it can easily be extended thanks to PROLOG flexibility.

All of the stages discussed above are supported by MIRELLA, an integrated CASE environment. In the following subsections the *guidelines* for each phase are given.

An example schematically showing the subsequent design phases is presented in Fig. 19.1.

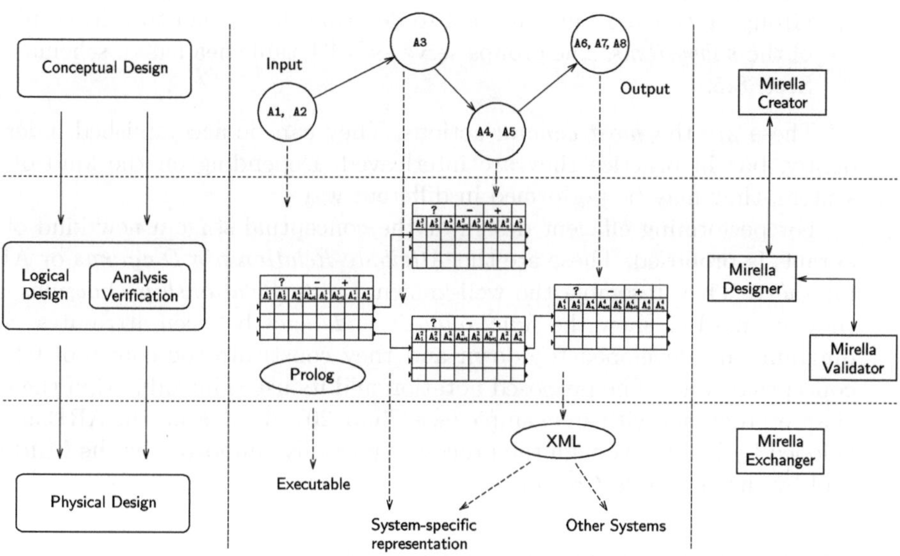

Fig. 19.1. Phases of the integrated design process

One of the most important features of this approach is the *separation of logical and physical design*, which also allows for a transparent, *hierarchical* design process.

19.3 Conceptual Design Phase with ARD Diagrams

The first phase of the system design process, the *conceptual design*, is the most abstract one. It is based on the *systems theory* approach.

It includes the following steps:

1. Defining the system *goal*, i.e. what the system is supposed to do in terms of its principal functionalities.
2. Identifying the system *inputs* and *outputs*, as well as their constraints.
3. Identifying what is needed to give a description of system *internal state* (or what information should be stored (memorized) inside the system) by defining the *conceptual variables* for representing system parameters at the abstract level.
4. Determining the functioning of the system, i.e. how the output is obtained for a given input and memory state, and how the memory state should change depending on the previous state and the input.
5. Identifying the internal system structure, along with the data and control flow.
6. Defining the system *attributes* and their *domains* — they describe system *inputs*, *outputs* and *state*.
7. Grouping the attributes needed to describe the system transition in each of the *subsystems*. The groups serve as XTT table headers or schemas, see Sect. 8.5.

These are the *most general* actions. They can be accomplished independently, but in practice they are interleaved. Depending on the kind of the system, they may be performed in different ways.

For performing efficient design at the conceptual stage a new kind of diagrams is proposed. These are the *Attribute-Relationship Diagrams* or ARD, for short. They resemble the well-known *Entity-Relationship Diagrams* and they are used to show the *functional dependencies* between attributes. Such diagrams are developed top-down and they constitute the core tool for the conceptual stage. The proposed notation and its use is introduced in the next chapter, together with an example (see Chap. 20). At present, the ARD are developed by hand, although the process is partially supported by the MIRELLA tool by the *Attribute Creator*.

19.3.1 Conceptual Modelling using ARD

The *conceptual design* of the RBS aims at modelling the most important features of the system, i.e. attributes and functional dependencies among them. During this design phase the ARD modelling method is used. ARD stands for *Attribute-Relationship Diagrams*. It allows for specification of functional dependencies of system attributes using a visual representation.

The concept of ARD constitutes a new idea, introduced for the purposes of this book; it was simultaneously presented in [100]. An attempt of formulation of ARD as a conceptual modelling tool for RBS is incorporated below.

The key underlying assumption in design of rule-based systems with knowledge specification in attributive logics is that, similarly as in the case of Relational Databases [23], the attributes are *functionally dependent*.

Let X, Y denote some sets of attributes, $X = \{X_1, X_2, \ldots, X_n\}$ and $Y = \{Y_1, Y_2, \ldots, Y_m\}$. We say that there is a functional dependency of Y from X, which is denoted as $X \to Y$, iff having established the values of attributes from X all the values of attributes of Y are defined in a unique way. Further, we are interested in the so-called *full* or *complete* functional dependencies, i.e. ones such that there is $X \to Y$ but for any proper subset $X' \subset X$ it is not true that $X' \to Y$; in other words, *all* the values of attributes of X are necessary to determine the values of Y.[2] A basic ARD table for specification of such a dependency is presented in Fig. 19.2.

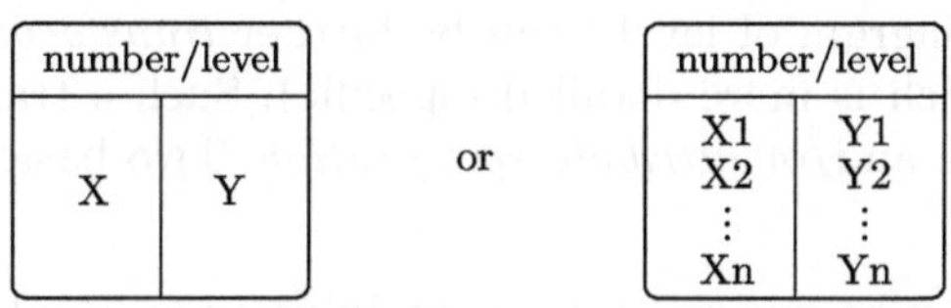

Fig. 19.2. An ARD table: the basic scheme for $X \to Y$

In the figure, the attributes on the left (i.e. the ones of X) are the independent ones, while the ones on the right (the ones of Y) are the dependent ones. The sets of variables (e.g. X or Y) will be referred to as *conceptual variables* and they are specified with a set of detailed, atomic attributes.

An ARD *diagram* is a conceptual system model a certain level of abstraction. It is composed of one or several ARD *tables*. If there are more than one ARD table, a *partial order* relation among the tables is represented with *arcs*. For intuition, this partial order means that attributes which are dependent in a preceding table must be established first, in order to be used as independent attributes in the following table (i.e. a partial order of determining attribute values is specified).

The ARD model is also a hierarchical model. The most abstract level 0 diagram shows the functional dependency of *input* and *output* system attributes. Lower level diagrams are less abstract, i.e. they are closer to full system specification. They contain also some intermediate conceptual variables and attributes.

As it was mentioned above, system attributes can be represented as *conceptual variables* on the abstract levels. The conceptual variables are further specified with one or more *physical attributes* on lower level (less abstract) diagrams. In the subsequent design stages physical attributes are directly mapped into logical structure of the system and implementation. At the final, most detailed specification no conceptual variables are allowed; all the attributes must represent the physical (atomic) attributes of the system.

[2] In case of a classical relational database table with the scheme specified with attributes $\{X_1, X_2, \ldots, X_n, Y_1, Y_2, \ldots, Y_m\}$ X will be considered as the *key*.

The table *heading* contains the table identifier which is a path-dot construction of the following recursive form:

- single number 0 is the label for the top-level single-table diagram,
- a sequence $0.[path].m$ is the label of the m-th table at the level $length(path)+1$, where $[path]$ is empty string, a single number or a sequence composed of numbers separated with dots.

In fact, such an identifier construction defines a tree, and a particular identifier specifies a path starting from the root and identifying a table component in a unique way.

An ARD diagram of level i can be further *transformed* into a diagram of level $i+1$, which is more detailed (specific). Such a transformation includes *table expansion* and/or *attribute specification.* Two basic *transformations* are considered:

1. *Horizontal split*, where a table containing conceptual variables is *expanded* into two tables, containing more detailed, intermediate conceptual variables or physical attributes,
2. *Vertical split*, where a table containing two (or more) dependent attributes is expanded into two (or more) independent tables.

During the transformation a conceptual variable can be specified (substituted) by more specific conceptual variables or a physical attribute, so that in the last, most detailed level diagram, only physical attributes are present.

The final, most detailed ARD must satisfy certain criteria. An ARD diagram can be considered *correct* iff it is consistent with the existing functional dependencies of the attributes and the specification enables determining all the values of (output) attributes once the input values are obtained. The correctness of the final, most detailed level is translated into the following conditions:

- all of the attributes are the physical attributes of the system (no conceptual variables are allowed),
- all of the input attributes (the independent attributes in the tables to which no arcs point to) are the system inputs (they are determined outside of the system),
- any ARD table specifies a full functional dependency (local correctness requirement), and
- for any path from the input to the output attributes no unestablishable attribute occurs, i.e. when traversing such a path from the left to the right, every attribute is either an input attribute or its first occurrence is as a dependent attribute in some ARD table.

The conditions above assure the possibility of determining all the values of all the attributes with respect to the functional dependencies among the attributes.

On the *machine readable level* ARD can be represented in an XML-based *ARDML* (*ARD Markup Language*) suitable for data exchange operations, as well as possibly transformations to other diagram formats.

19.3.2 Attributes Definition with the Attribute Creator

The *Attribute Creator* allows for defining system attributes and domain constraints. An example is shown in Fig. 19.3. Furthermore, using the defined attributes, the process of building XTT tables is supported.

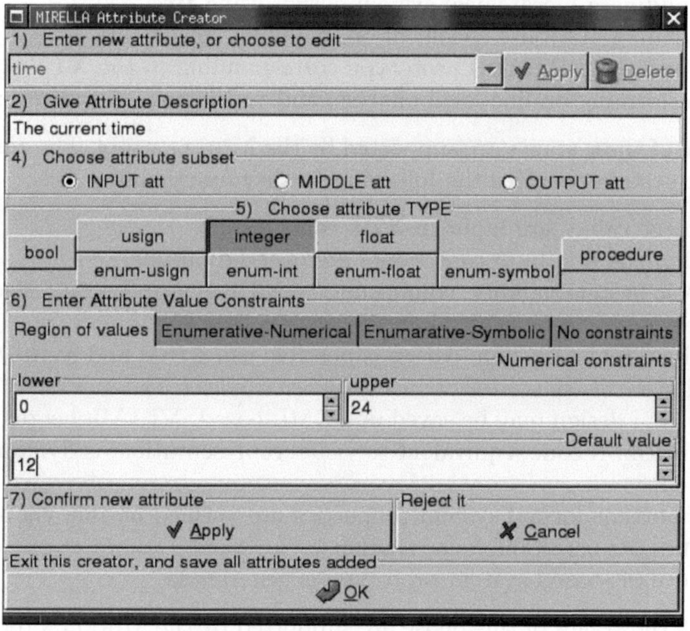

Fig. 19.3. Attribute creator

To summarize this phase it may be concluded that:

- the *input* of this phase is the general description of the system;
- the *output* of this phase is:
  - an ARD *conceptual diagram* describing: input, output as well as main system components and dependencies between them (this indicates partially the the control strategy to be implemented);
  - specification of system *attributes*, and *attribute domains*;
  - design of attribute groups — XTT table headers.

The creation of the ARD conceptual diagram will be shown in Chap. 20 with the use of the thermostat example.

19.4 Logical Design Phase with XTT

The *logical design phase* uses the results of the conceptual design.

It involves the following stages:

1. Specifying the full headers of the tables with appropriate attributes, i.e. defining the structure of the XTT.
2. Defining the rules through filling the XTT tables with particular values. Each table specifies the system behavior in a certain context.
3. Linking tables together with XTT connections, thus specifying inter-table relations.
4. Building a hierarchical system design using *tree* facilities.
5. Refining attribute specification including constraints.
6. Generating PROLOG prototype corresponding to the XTT structure.
7. Performing on-line local analysis and verification of system properties.

All of these stages are supported by the MIRELLA tool. The current version of the system provides the following further functionality:

- object values are input in XTT cells;
- all of the attribute constraints are checked on-line;
- basic system validity, conforming to XTT syntax is controlled during the design;
- a hierarchical system design is possible using *tree*, and *context tables* facilities;
- system design may be saved in a XML-based, XTTML language (see [92]);
- a PROLOG code, equivalent to visual representation, is dynamically generated;
- important formal system properties are verified on-line via an embedded PROLOG-based inference engine;
- PROLOG-based system prototype is generated.

All the stages of this phase are supported by the MIRELLA tool. Particular examples are given in Chap. 20; more examples are in [92].

To summarize this phase it may be concluded that:

- the *input* of this phase is the output of *conceptual design*;
- the *output* of this phase is:
  - a full XTT-based system design,
  - complete PROLOG representation of the above,
  - system description in XML-based format,
  - system analysis reports and visualization of results.

This is the main design phase. The design process is hierarchical. The presented stages can be schematically illustrated as in Fig. 19.4.

It is important that the proposed approach allows for verification and optimization during the stage of logical design. In fact, the issues of system

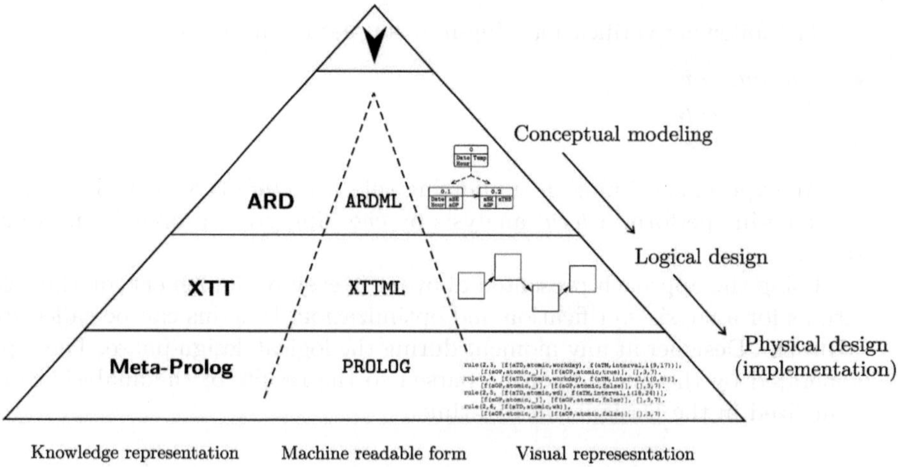

Fig. 19.4. Phases of the integrated design process

analysis, *optimization* and *verification* play an important role in the integrated design process. The integrated framework for early system analysis is presented in the following section.

19.5 The Analysis and Verification Framework

During the logical design phase incremental rule-base synthesis is supported by the on-line, PROLOG-based system analysis and verification framework. The *framework* allows for the implementation of different external verification and analysis modules (plug-ins). Examples of such modules were presented in the former part of this book devoted to verification problems. The framework can be integrated with the XTT inference engine described in Sect. 20.

The external analysis, verification and optimization modules are implemented in PROLOG. They have direct access to the system knowledge base. Each module reads the XTT rule base and performs the analysis of the given property and produces a report. The report can be optionally visualized in the MIRELLA Designer. Since the modules have the access to the PROLOG-based XTT system description, it is also possible to implement dynamic rule correction algorithms.

Building a complete rule-based system analysis and verification suite is a complex problem. It could be solved in many alternative ways. Several illustrative examples of analysis and verification plug-ins have been discussed in Part III of this book.

The following verification plug-in examples are available:

- *subsumption*,
- *determinism*,
- *completeness.*

An experimental plug-in supporting rule *reduction* is also available. All of the plug-ins perform a *local* analysis by checking given properties in a given table.

Using the approach presented above it is easy to implement another algorithms for analysis, verification and optimization. Plug-ins can be called from MIRELLA Designer at any moment during the logical design phase. The report generated by the plug-in can be parsed so the results of the analysis can be visualized in the Designer in real-time.

19.6 Implementation Phase

The final phase of the design process is the *physical design*, or *implementation.*

In this phase the following stages can be identified:

1. Testing the prototype via an inference engine shell.
2. Debugging the prototype by built-in SWI-PROLOG both text-based and visual tracer.
3. Generating stand-alone application prototype (or a rule-base to be used with a specific rule interpreter).

These stages are described in more detail in the following subsections.

19.6.1 Testing the Prototype

Once the PROLOG code has been generated it can be tested using the inference engine shell provided with MIRELLA (see [92]); for an example implementation see also the next section. The shell allows to input the initial data, which can also be read from *fact tables* (see [92]). The system prototype can then be executed in the shell. The examples are given in [92], where practical rule-based systems implementations are studied. During the testing process the code can also be debugged.

19.6.2 Debugging the Prototype

One of the reasons SWI-PROLOG [138] has been chosen for MIRELLA is its advanced features when it comes to debugging. SWI-PROLOG has a 6-port tracer, which extends the standard 4-port tracer described in [19] with two additional ports. The optional *unify port* allows the user to inspect the result after unification of the head. The *exception port* shows exceptions raised by *throw/1* or one of the built-in predicates.

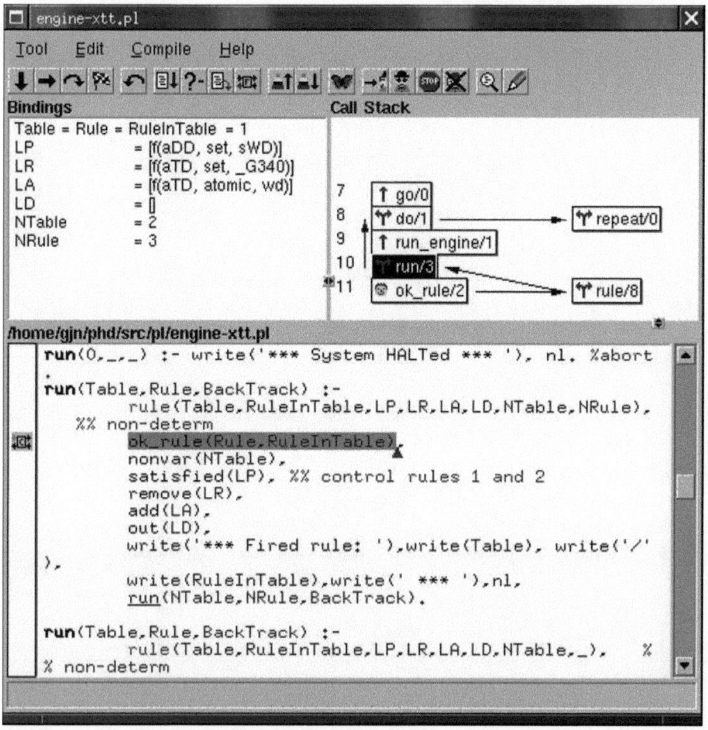

Fig. 19.5. Graphical PROLOG tracer

The code can also be *profiled* by the built-in profiler which displays call and time statistics of the program. It can also be *visually* traced by the *graphical tracer*; an example session is showed in Fig. 19.5. This stage is typical for general PROLOG code debugging.

19.6.3 Generating Stand-Alone Application

Finally, a stand-alone prototype can be generated using SWI compiler (see Chapt. 9 in [138]). In this stage three components are *linked* together: the XTT *inference engine*, the system *rule base*, the SWI *runtime*. Alternatively, an independent rule base can be produced.

This approach allows for building stand-alone expert system applications, running *independently* of MIRELLA, or SWI–PROLOG environment on *multiple platforms*, including: GNU/Linux, Unix, Windows, MacOS. Since GNU/Linux is suitable for building *embedded systems* [90, 91], the approach presented here could eventually allow for building *embedded rule-based systems*.

20

Design Example: Thermostat

In this chapter we return once again to the thermostat example rule-based system [102] (see Sect. 8.6). Using this example it is shown how the design approach based on the use of the XTT works and what the results are. We end up with the PROLOG implementation of this system.

The example is presented with some details paid to the following elements:

- *Description* — the system is briefly introduced.
- *Rules* — system rule-base in the original rule-based form is introduced (control specification).
- *Attributes* — the specification of system attributes, domains and constraints is given.
- *Attribute-Relationship Diagrams* — the conceptual design with use of the ARD is presented.
- *Logical design* — logical system design in XTT is presented.
- *Physical design* — PROLOG code representing the logical structure generated by MIRELLA is shown.
- *Implementation* — example of PROLOG code execution in the inference engine is presented.
- *Analysis and Remarks* — results of system analysis (if applicable) and observations made during the process of system design are given.

The first four stages are parts of the conceptual design. The second stage — standard rule-base — is included in order to provide textual specification of the control task; it takes the form of rules, however, in general, it can be in any syntactic form of text. In more complex problems there will be no rule-like specification, and the attributes, their values and domains, and dependencies among attributes will have to be identified from the accessible specification and knowledge of the domain experts.

The presented example illustrates design stages and the use of introduced concepts. It exposes possible improvements in terms of knowledge representation density, as well as the design process itself.

A. Ligęza: *Logical Foundations for Rule-Based Systems*, Studies in Computational Intelligence (SCI) **11**, 263–275 (2006)
www.springerlink.com

20.1 Thermostat Control System

Description

Let us consider a simple but illustrative rule-based control system for setting the required temperature in a room, depending on the day, season, hours, etc. The example is based on Thermostat example found in [102][1].

The goal of the system is to set a temperature at a certain *set point*, which is the output of the system. The input is the current time and date. The temperature is set depending on the particular part of the week, season, and working hours.

Rules

The original ([102]) Thermostat system rule base follows:[2]

Rule 1: **if** the day is Monday **or** the day is Tuesday **or** the day is Wednesday **or** the day is Thursday **or** the day is Friday **then** today is a workday.

Rule 2: **if** the day is Saturday **or** the day is Sunday **then** today is the weekend.

Rule 3: **if** today is workday **and** the time is 'between 9 am and 5 pm' **then** operation is 'during business hours'.

Rule 4: **if** today is workday **and** the time is 'before 9 am' **then** operation is 'not during business hours'.

Rule 5: **if** today is workday **and** the time is 'after 5 pm' **then** operation is 'not during business hours'.

Rule 6: **if** today is weekend **then** operation is 'not during business hours'.

Rule 7: **if** the month is January **or** the month is February **or** the month is December **then** the season is summer.

Rule 8: **if** the month is March **or** the month is April **or** the month is May **then** the season is autumn.

Rule 9: **if** the month is June **or** the month is July **or** the month is August **then** the season is winter.

Rule 10: **if** the month is September **or** the month is October **or** the month is November **then** the season is spring.

Rule 11: **if** the season is spring **and** operation is 'during business hours' **then** thermostat_setting is '20 degrees'.

---

[1] An appropriate knowledge base is given in [102], pages 41–43; for convenience we list it in textual form here. Note that, from the point of view of *control theory* the specification is not one of a *thermostat* system (which has as a task temperature stabilization at a certain *set point*), but it is a specification of *set point* selection algorithm of a higher-level (adaptation) controller.

[2] One can be confused when looking at rules 7–10 defining *season*. They look strange only if one lives on the northern hemisphere. However, the author of [102] lives in Australia, which explains the 'confusion'.

Rule 12: **if** the season is spring **and** operation is 'not during business hours' **then** thermostat_setting is '15 degrees'.
Rule 13: **if** the season is summer **and** operation is 'during business hours' **then** thermostat_setting is '24 degrees'.
Rule 14: **if** the season is summer **and** operation is 'not during business hours' **then** thermostat_setting is '27 degrees'.
Rule 15: **if** the season is autumn **and** operation is 'during business hours' **then** thermostat_setting is '20 degrees'.
Rule 16: **if** the season is autumn **and** operation is 'not during business hours' **then** thermostat_setting is '16 degrees'.
Rule 17: **if** the season is winter **and** operation is 'during business hours' **then** thermostat_setting is '18 degrees'.
Rule 18: **if** the season is winter **and** operation is 'not during business hours' **then** thermostat_setting is '14 degrees'.

There are certain patterns that may be observed in the rule-base. They will be used in the following section containing the corresponding XTT structure.

Attributes

By analyzing the above textual specification of the desired operation of the system the attributes can be identified and their domains can be determined. The original basic list of attributes and their possible values are also given in [102]. The attributes can be extracted from informal specification of the rules presented in Sect. 8.6 or directly from the rules specified in natural language, provided that this was the form of initial specification of the system to be designed. They are shown in Table 20.1. The names of the attributes are self-explanatory.

Considering system input, state and output, as well as rule-base, a formal attribute specification, conforming to XTT method is shown in Table 20.2.

When comparing Tables 20.1 and 20.2, several design decisions can be observed:

- *time* attribute has been defined as `integer` in order to allow arithmetic comparison,
- *operation* attribute has been defined as `boolean`, due to its true/false (boolean) use,
- *today*, *season*, *operation* attributes have been defined as 'middle' attributes, since their values are inferred from input attribute values.

Let us show the use of *Attribute-Relationship Diagram* for designing the conceptual structure of the system.

On the base of analysis of the textual specification of the rules one can notice that the temperature setting (to be denoted by `Temp` can be established if only current date (`Date`) and hour (`Hour`) are known. It means that there is functional dependency of the form

$$Date, Hour \rightarrow Temp.$$

Table 20.1. Thermostat attributes

Name	Symbol	Values
day	*aDD*	`Monday, Tuesday, Wednesday, Thursday, Friday, Saturday, Sunday`
today	*aTD*	`workday, weekend`
time	*aTM*	`between 9am and 5pm, before 9am, after 5pm`
month	*aMO*	`January, February, March, April, May, June, July, August, September, October, November, December`
season	*aSE*	`summer, autumn, winter, spring`
operation	*aOP*	`during business hours, not during business hours`
thermostat	*aTHS*	`14, 15, 16, 18, 20, 24, 27`

Table 20.2. XTT Thermostat attributes specification

Name	Subset	Type	Constraints
day	input	enumerative, symbolic	{`monday, tuesday, wednesday, thursday, friday, saturday, sunday`}
time	input	integer	⟨ `0,24` ⟩
month	input	enumerative, symbolic	{`january, february, march, april, may, june, july, august, september, october, november, december`}
today	middle	enumerative, symbolic	{`workday, weekend`}
season	middle	enumerative, symbolic	{`spring, summer, autumn, winter`}
operation	middle	boolean	—
thermostat	output	enumerative, integer	{`14, 15, 16, 18, 20, 24, 27`}

This can be denoted with the following diagram presented in Fig. 20.1.

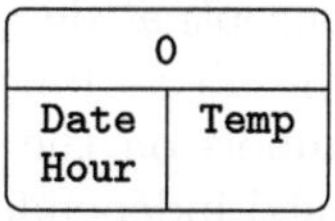

Fig. 20.1. The ARD diagram of level 0

The meaning of the diagram is as follows: at the most abstract level of specification (heading: 0), the temperature (`Temp`, the right column) depends on date and hour (`Date` and `Hour`, the left column). The names `Date`, `Hour` and `Temp` are *conceptual variables* and they will be further specified with one or more attributes.

The next step of the design consist of horizontal splitting of the top level ARD — new, intermediate variables are introduced showing the operation of the system in a more precise way. From the provided specification it can be noticed that the temperature is determined by season and operation, and both the attributes can be determined from date and hour. Hence, the following functional dependencies hold:

$$aSE, aOP \rightarrow aTHS$$

and

$$Date, Hour \rightarrow aSE, aOP.$$

The next diagram is shown in Fig. 20.2.

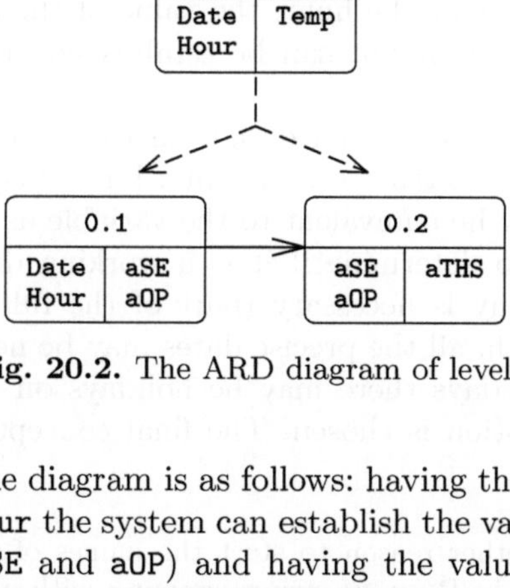

Fig. 20.2. The ARD diagram of level 1

The meaning of the diagram is as follows: having the values of conceptual variables `Date` and `Hour` the system can establish the values of season and operation attributes (`aSE` and `aOP`) and having the values of these attributes it can further establish the temperature, this time denoted as `aTHS`. The convention is that final attributes are denoted as aATTRIBUTE to indicate that they are single attributes used in the logical and physical implementation[3].

A further step of the conceptual design consists of *vertical split* which can be performed since in the above functional dependency $Date, Hour \rightarrow aSE, aOP$ there are two symbols on the right-hand side. The value of the season (aSE) can be determined independently from operation (aOP). This is shown in Fig. 20.3.

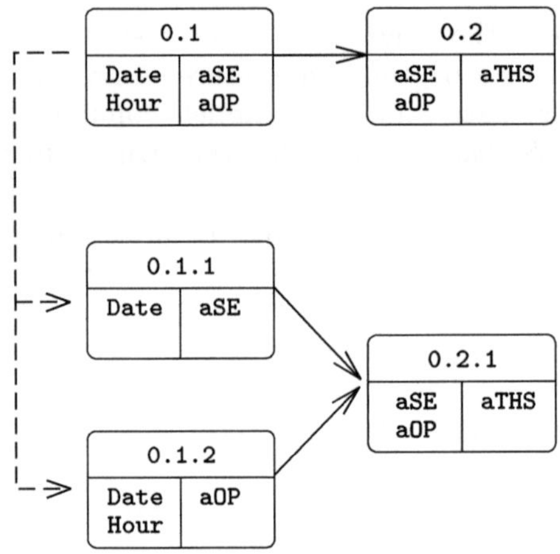

Fig. 20.3. The ARD diagram of level 2

A further design step consists of another *horizontal split* operation. Having the date, the system can establish whether it is a working day or a weekend which is denoted with attribute aTD. Further, depending on that knowledge and taking into account the hour, the value of the aOP attribute denoting if this is operation time or not can be established. This split is shown in the next Fig. 20.4.

To complete the conceptual design all the conceptual variables must be replaced with actual attributes; such attributes may be formed as a part of the variable or may be equivalent to the variable as a whole.

For example, to determine if it is a working day or not, basically only the name of the day is necessary (part of the full date); in another, more elaborated approach, all the precise dates may be necessary, since apart from Saturdays and Sundays there may be holidays on some other days as well. Below, the first option is chosen. The final conceptual diagram is shown in Fig. 20.5.

[3] There is also another reason to start the names of the attributes will a lower case letter 'a' — in PROLOG strings starting with an upper case letter denote variables. Further, the first letter denotes the role of the name, i.e. 'a' stands for an attribute, 's' denotes a set while 'i' — an interval.

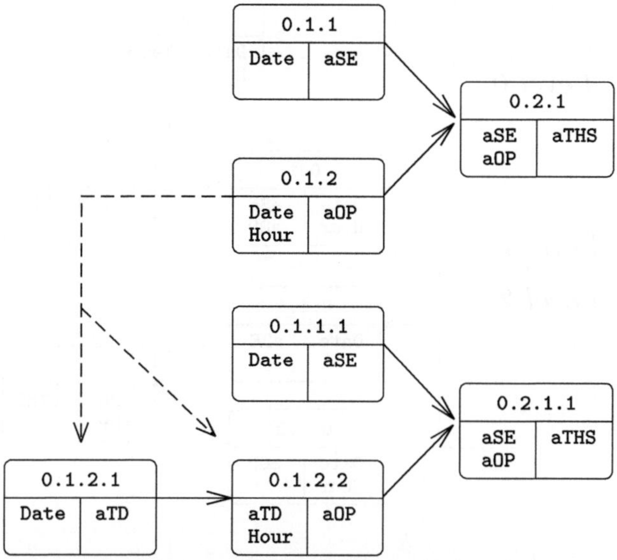

Fig. 20.4. The ARD diagram of level 3

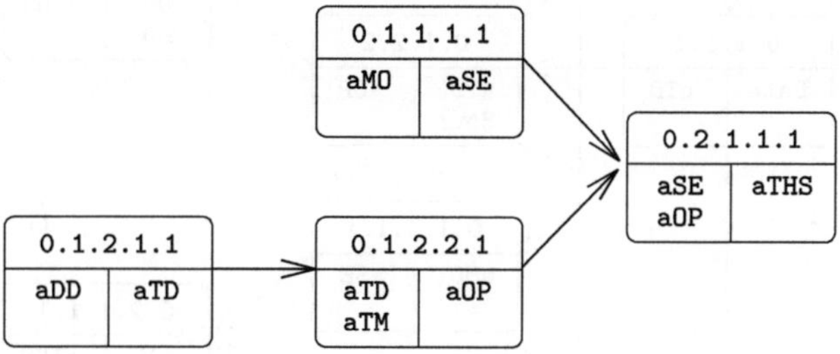

Fig. 20.5. The ARD diagram of level 4

In the above presentation the arrows show the partial order relation defining the order of inference. The dashed lines show how the split is performed.

Note also the accepted convention for enumerating the components. As the design is performed top-down, any component keeps the number of its parent node completed with its individual number — this is achieved by using the path-dot expressions. Simultaneously, the number of positions indicates the level of detail.

Note that in the final diagram a partial order relation among components, following from the functional dependencies identified, is represented with the use of arrows. This is an important piece of information and it will be further used for specifying efficient control execution for the rules.

To summarize, at the end of the conceptual design stage the complete ARD diagram is obtained (see Fig. 20.6).

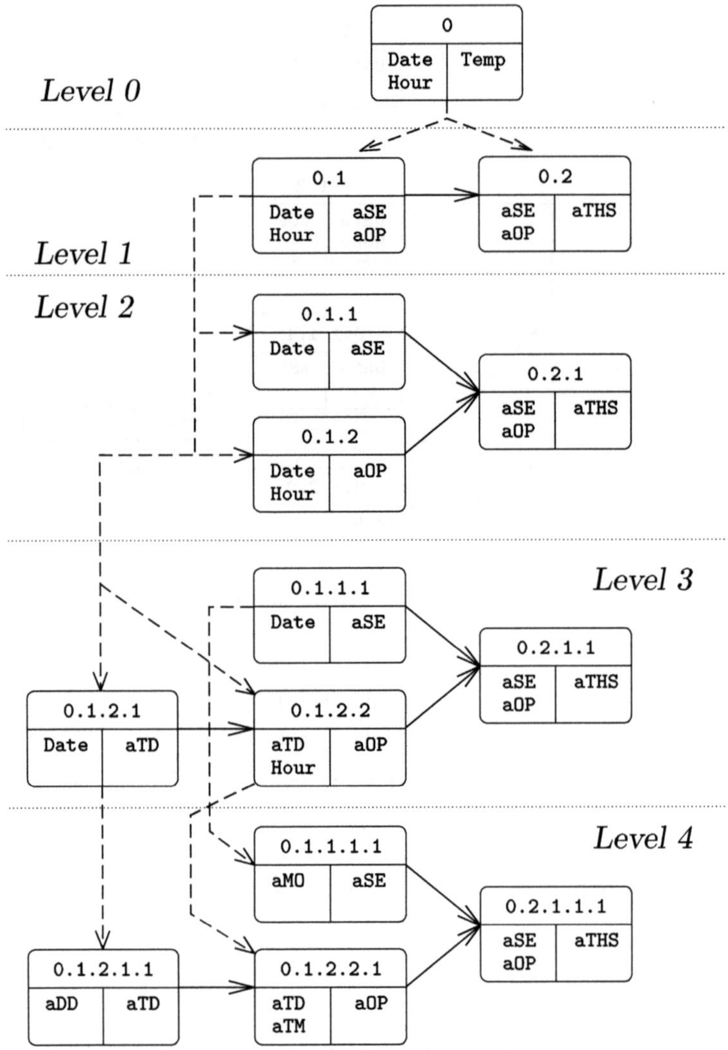

Fig. 20.6. The complete ARD diagram for levels 0 to 4

Each of the lowest-level components will be used to specify the scheme of an XTT table in the logical design stage. The left-hand attributes form the precondition part, while the right-hand ones refer to conclusion; it can be further classified as final conclusion or assert (retract) attributes. In the presented example four XTT tables are identified.

Logical Design

In the presented example the rules can be divided in a natural way into groups producing the same kind of decision, where the precise decision depends on precise preconditions. Furthermore, in the groups the preconditions of the rules employ the same linguistic variables, but different values. Hence, the XTT tables can be formed directly from rules.

In general, the specification does not necessarily take the form of such transparent rules. In such cases the use of ARD diagrams greatly simplifies the design, and especially the identification of functional dependencies among attributes.

As follows from the conceptual design stage and the obtained ARD diagram, four different XTT tables will have to be designed. The design can be performed directly on the base of identification of four groups of rules defining decisions whether *today* is `workday` or `weekend` (rules 1 and 2), whether the *time* is during business hours or not (rules 3–6), deciding what season do we have (rules 7–10), and finally indicating the setting of the thermostat (rules 11–18).

For each group a separate XTT table can be built — this seems to be a natural and efficient approach. Since the rules use the same variables, the specification of attributes in the table is given just once and all the columns are necessary for any rule in a table.

The logical design of the tables is presented in Sect. 8.6. The tables are defined with the graphical editor tool of MIRELLA. This results in creating the following tables:

- **today** table infers the value of *today* attribute, it is a *root table*;
- **operation** table checks the time of the day;
- **season** table infers the current *season*;
- **temperature** table makes the decision about thermostat temperature.

For continuing presentation of the example, below the tables imported from Chap. 8 are recalled.

Table 20.3. Context 1: none. ARD component No.: 0.1.2.1

Info	Prec	Retract	Assert	Decision	Ctrl	
I	*aDD*	*aTD*	*aTD*	*H*	*N*	*E*
1	*sWD*	—	*wd*		2.3	1.2
2	*sWK*	—	*wk*		2.6	1.1

Table 20.4. Context 2: $aTD \in \{wd, wk\}$. ARD component No.: 0.1.2.2

Info	Prec		Retract	Assert	Decision	Ctrl	
I	*aTD*	*aTM*	*aOP*	*aOP*	*H*	*N*	*E*
3	*wd*	[9:00, 17:00]	—	*dbh*		3.7	2.4
4	*wd*	[00:00, 09:00]	—	*ndbh*		3.7	2.5
5	*wd*	[17:00, 24:00]	—	*ndbh*		3.7	2.6
6	*wk*	–	—	*ndbh*		3.7	2.3

Table 20.5. Context 3: none. ARD component No.: 0.1.1.1

Info	Prec	Retract	Assert	Decision	Ctrl	
I	*aMO*	*aSE*	*aSE*	*H*	*N*	*E*
7	*sSUM*	—	*sum*		4.13	3.8
8	*sAUT*	—	*aut*		4.15	3.9
9	*sWIN*	—	*win*		4.17	3.10
10	*sSPR*	—	*spr*		4.11	3.7

Table 20.6. Context 4: $aSE \in \{spr, sum, aut, win\} \wedge aOP \in \{dbh, ndbh\}$. ARD component No.: 0.2.1.1

Info	Prec		Retract	Assert	Decision	Ctrl	
I	*aSE*	*aOP*			*aTHS*	*N*	*E*
11	*spr*	*dbh*			20	1.1	4.12
12	*spr*	*ndbh*			15	1.1	4.13
13	*sum*	*dbh*			24	1.1	4.14
14	*sum*	*ndbh*			17	1.1	4.15
15	*aut*	*dbh*			20	1.1	4.16
16	*aut*	*ndbh*			16	1.1	4.17
17	*win*	*dbh*			18	1.1	4.18
18	*win*	*ndbh*			14	1.1	1.1

The tables have been designed on the base of the textual specification of rules. The control part of the tables has been designed with taking into account the partial order determined with ARD and so that the system works in a robust and efficient way. Specification of the retract part with the underscore means in fact that any value of aTD should be removed (the variables should be cleared). The same convention is applied in further tables. More details have been presented in Chap. 8.

The XTT structure for the thermostat system has been designed using MIRELLA Designer. All of the tables and connections between them are shown in Fig. 20.7.

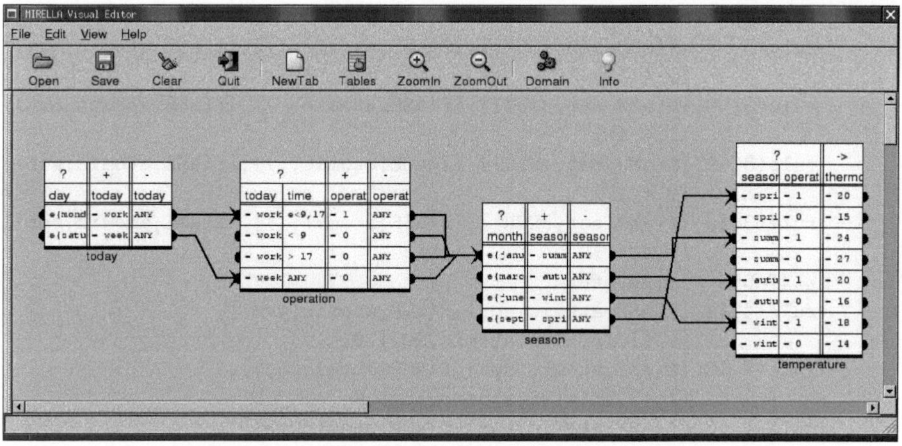

Fig. 20.7. Thermostat system design in MIRELLA

The order of execution was introduced in an arbitrary way, so that it is consistent with the partial order specified with the ARD diagram presented in Fig. 20.5.

Physical Design

The following PROLOG rule-base has been generated by MIRELLA:

```
set(sWD,[monday,tuesday,wednesday,thursday,friday]).
set(sWK,[saturday,sunday]).
set(sSUM,[january,february,december]).
set(sAUT,[march,april,may]).
set(sWIN,[june,july,august]).
set(sSPR,[september,october,november]).

%%%Rulebase
%%%Table: today
rule(1,1,[f(aDD,set,sWD)],[f(aTD,set,_)],[f(aTD,atomic,wd)],
        [],2,3).
rule(1,2,[f(aDD,set,sWK)],[f(aTD,set,_)],[f(aTD,atomic,wk)],
        [],2,6).
%%%Table: operation
rule(2,3,[f(aTD,atomic,wd),f(aTM,interval,i(9,17))],
        [f(aOP,atomic,_)],[f(aOP,atomic,yes)],[],3,7).
rule(2,4,[f(aTD,atomic,wd),f(aTM,interval,i(0,8))],
        [f(aOP,atomic,_)],[f(aOP,atomic,false)],[],3,7).
rule(2,5,[f(aTD,atomic,wd),f(aTM,interval,i(18,24))],
        [f(aOP,atomic,_)],[f(aOP,atomic,false)],[],3,7).
rule(2,6,[f(aTD,atomic,wk)],[f(aOP,atomic,_)],[f(aOP,atomic,no)],
        [],3,7).
```

```
%%%Table: season
rule(3, 7,[f(aMO,set,sSUM)],[f(aSE,atomic,_)],[f(aSE,atomic,sum)],
          [],4,13).
rule(3, 8,[f(aMO,set,sAUT)],[f(aSE,atomic,_)],[f(aSE,atomic,aut)],
          [],4,15).
rule(3, 9,[f(aMO,set,sWIN)],[f(aSE,atomic,_)],[f(aSE,atomic,win)],
          [],4,17).
rule(3,10,[f(aMO,set,sSPR)],[f(aSE,atomic,_)],[f(aSE,atomic,spr)],
          [],4,11).
%%%Table: temperature
rule(4,11,[f(aSE,atomic,spr),f(aOP,atomic,yes)],
          [],[],[f(aTHS,atomic,20)],0,_).
rule(4,12,[f(aSE,atomic,spr),f(aOP,atomic,no)],
          [],[],[f(aTHS,atomic,15)],0,_).
rule(4,13,[f(aSE,atomic,sum),f(aOP,atomic,yes)],
          [],[],[f(aTHS,atomic,24)],0,_).
rule(4,14,[f(aSE,atomic,sum),f(aOP,atomic,no)],
          [],[],[f(aTHS,atomic,17)],0,_).
rule(4,15,[f(aSE,atomic,aut),f(aOP,atomic,yes)],
          [],[],[f(aTHS,atomic,20)],0,_).
rule(4,16,[f(aSE,atomic,aut),f(aOP,atomic,no)],
          [],[],[f(aTHS,atomic,16)],0,_).
rule(4,17,[f(aSE,atomic,win),f(aOP,atomic,yes)],
          [],[],[f(aTHS,atomic,18)],0,_).
rule(4,18,[f(aSE,atomic,win),f(aOP,atomic,no)],
          [],[],[f(aTHS,atomic,14)],0,_).
```

The rule base presented above uses the rule format defined in the Sect. 13.3.

The rules of the last table (temperature) have modified the Next and Else part — 0 means that the system stops after execution of a single cycle.

Implementation

An example execution, using the XTT inference engine is shown below. The shell also serves as an environment for an *interactive simulation.*

```
Welcome to SWI-Prolog (Multi-threaded, Version 5.2.6)
?- [engine-xtt].
% engine-xtt compiled 0.00 sec, 9,784 bytes
Yes
?- go.
--- This is the XTT Prolog shell. ---
--- Enter "load", "run", "runbt", "quit" at the prompt. ---
>>> load.
--- Clearing old rule base ---
--- Enter rule file name: >>> therm-xtt.
% therm-xtt compiled 0.00 sec, 7,208 bytes
>>> run.
--- Running Engine ---
```

```
*** Welcome to the Thermostat Rule-Base System ***
*** Provide the initial data ***
Month: [january--december]: >>> march.
Day:   [monday--sunday]:    >>> friday.
Hour:  [0--24]:             >>> 16.
+++ No output +++
*** Fired rule: 1/1 ***
+++ No output +++
*** Fired rule: 2/3 ***
+++ No output +++
*** Fired rule: 3/8 ***
+++ Rule out: f(aTHS, atomic, 20) +++
*** Fired rule: 4/15 ***
*** System HALTED ***
```

Using the initial data the system produces the decision (`Rule out` lines), by firing a sequence of rules (`Fired rule` lines).

Analysis and Remarks

The specification of the system with the use of the XTT method seems to be both concise and easy to analyze. The original Thermostat example is specified in such way that the system has basic properties such as determinism, completeness and lack of subsumption preserved.

However, the example reduction plug-in, provided with MIRELLA, is able to detect a possible *reduction*:

```
?- vpr(4).
*** Rule: 4.11 may be glued with rule: 4.15
    reduced fact: f(aSE, set, [spr, aut])
*** Rule: 4.15 may be glued with rule: 4.11
    reduced fact: f(aSE, set, [aut, spr])
No more reduction of rules in table 4
```

The following rules:

```
rule(4,11,[f(aSE,atomic,spr),f(aOP,atomic,yes)],
          [],[],[f(aTHS,atomic,20)],0,_).
rule(4,15,[f(aSE,atomic,aut),f(aOP,atomic,yes)],
          [],[],[f(aTHS,atomic,20)],0,_).
```

could be substituted by a single rule with use of a *non-atomic* value:

```
rule(4,11,[f(aSE,set,[aut, spr]),f(aOP,atomic,yes)],
          [],[],[f(aTHS,atomic,20)],0,_).
```

The XTT approach allowed for dividing the system knowledge base into four interconnected modules. Non-atomic attribute values proved to be useful in specifying precondition attribute values and in table reduction.

21

Concluding Remarks

The book presents logical foundations for rule-based systems, as seen by the Author. An attempt has been made to provide an in-depth discussion of logical and other aspects of such systems, including languages for knowledge representation, inference mechanisms, inference control, design and verification. The ultimate goal was to provide a deeper theoretical insight into the nature of rule-based systems and put together the most complete presentation including details so frequently skipped in typical textbooks.

The book is divided into four main parts, each of them further divided into several chapters.

The main parts present material on:

- logical foundations of rule-based systems (Part I);
- principles of rule-based systems structures, knowledge representation languages, inference and inference control (Part II);
- verification of formal properties of rule-based systems (Part III);
- design methodology for efficient development of such systems (Part IV).

Since '*thinking in terms of facts and rules is perhaps one of the most common ways of approaching problem definition and problem solving both in everyday life and under more formal circumstances*' the book may be useful to potentially wide audience, but it is aimed at providing specific knowledge for graduate, post-graduate and Ph.D. students, as well as knowledge engineers and research workers involved in the domain of AI. It also constitutes a summary of the Author's research and experience gathered through several years of his research work.

As mentioned in the Introduction, this book addresses the methodology of rule-based systems in a relatively complete and perhaps a bit complex, formal way. The main aim was to present the rule-based systems from logical perspective as viewed by the Author. Certain Author's concepts concerning rule-based systems were described in detail. Although the primary concern of this book may seem to be well-explored in the domain literature, both the structure and the contents of the book attempted at keeping individual,

A. Ligęza: *Logical Foundations for Rule-Based Systems*, Studies in Computational Intelligence (SCI) **11**, 277–279 (2006)
www.springerlink.com

Author-shaped character, an express personal experience of both theoretical and practical nature.

The concept of the book was to present in a single volume a spectrum of knowledge concerning rule-based systems, as understood in knowledge engineering, but also with going into details uncovered by other books on that topic. The book covers areas such as: logical foundations of rule-based systems (including knowledge representation and inference with propositional, attribute-based and first-order logic), knowledge representation, inference and inference control in rule-based systems, including extended forms of rules and specialized inference control mechanisms, definitions and verification of formal properties of rule-based systems assuring the correct work of them and finally design issues covering systematic design approach combined with on-line verification.

Some new concepts introduced in this book range from theoretical issues such as *backward dual resolution* as a tool for analysis of rule-based systems (especially for verification of completeness and reduction through gluing of rules) to practical design-supporting tools such as:

- *Attribute-Relationship Diagrams* (ARD),
- *eXtended Tabular Trees* (XTT).

A number of detailed problems were presented in formal way, analyzed and explained.

The concept of tabular systems discussed in this book seems to provide a new quality in knowledge representation. Simultaneously, it constitutes a step on the way to the postulated *algebraization* of knowledge. It also incorporates possibility of *hierarchical* knowledge representation and *hierarchical development* of a rule-based systems. Finally, it enables interleaving the stages of *verification* and *design*, so that a possibly correct system is designed and developed. The proposed experimental tool integrates these features in a single system and enables far going support of development of rule-based systems.

A dream of the Author would also be that this book serves as a source of inspiration for further research in the domain of rule-based systems and in AI as a whole. Although no new research directions have been explicitly defined in the book, in several places a certain kind of problems and barriers on one hand, and promises on the other, can be found between the printed lines of the text.

It seems that the following research possibilities open a wide and attractive research area for:

- development of the visual design procedures and development of supporting tools [92, 93, 94, 98], especially ones incorporating capabilities of reusability, pre-defined components, optimization of the designed code and its reduction [75, 79, 80], library of cases and case-based reasoning tools [143], verification [95], etc.;

- further automatization of construction of rule-based systems, especially combining development of such systems with learning and automatic knowledge acquisition techniques [18];
- development of 'intelligent documents' incorporating rule-based component for knowledge specification and execution and capable of automatic knowledge processing [76, 77];
- application of the rule-based system methods and technology to knowledge management [76, 96, 99];
- development of more elaborated knowledge representation and processing formalism, e.g. ones based on the concept of granularity, and further algebraization of knowledge processing [72, 73, 74, 77, 78];
- extension of the knowledge-representation formalism towards covering uncertainty [13].

To conclude, the book is aimed at presenting specific knowledge but simultaneously inspiring further research. It is not a closed, well structured and packed, textbook in its final form. Please find it open and inspiring, try to see the problems emerging from the presented solutions. Critical and creative thinking is always a necessary component of progress.

Part V

Closing Remarks and Appendices

A

Selected Rule-Based Systems and Tools

In this appendix a list of some selected rule-based systems and tools is provided[1]. It is aimed as an introductory information on some systems belonging to the class of AI tools incorporating rule-based technology.

Selected application tools are mentioned and some examples are pointed to. They refer to some experimental tools, developed as a result of theoretical research, and present some relatively new, distinguished features as well as the established standards. For more 'classical' examples one can also look into [51] and [48, 81, 113, 122, 128, 129].

A.1 Related Work and Knowledge Verification Tools

A.1.1 Kheops System

KHEOPS [42] is a real-time rule-based system shell. It has a reactive, forward-chaining interpreter. It is oriented toward time-critical, on-line applications. Its distinctive features include the compilation of a rule-base to the form of a specific decision tree which allows for checking some formal properties. However, it has a poor user interface and lacks support for an interactive system design.

KHEOPS is an advanced rule-base real time system. Its working idea is relatively simply: it constitutes a reactive, forward interpreter. However, it is relatively fast (response time can be below 15 milliseconds) and oriented toward time-critical, on-line applications. Its distinctive features include compilation of the rule-base to the form of specific decision tree which allows for checking some formal properties (e.g. completeness) and allows for evaluation of response time, dealing with time representation and temporal inference, and incorporation of specialized forms of rules, including universal quantification and C-expression. A detailed description of KHEOPS can be found in [42].

[1] The list is provided courtesy of Grzegorz Jacek Nalepa, Ph.D. and it is mostly based on an extract from his Ph.D. Thesis [92].

A. Ligęza: *Logical Foundations for Rule-Based Systems*, Studies in Computational Intelligence (SCI) **11**, 283–288 (2006)
www.springerlink.com

A.1.2 Prologa

PROLOGA is an interactive design tool for computer supported construction and manipulation of decision tables [133]. It is oriented towards decision-tables based knowledge representation. It provides a design environment along with some additional knowledge acquisition facilities. It allows for table verification and detection of common anomalies such as the lack of consistency, or redundant rules.

However, its knowledge representation method is limited to classic decision tables only. Its expressive power is limited to the propositional-calculus-based knowledge representation. The design support does not include a visual knowledge representation, only a simple graphical table representation.

A.1.3 KbBuilder

Although the principal idea to include the *verification* stage into the *design* process, as well as to support the design with flexible graphical environment of the CAD/CASE type dates back to [58, 59], there exist only few papers devoted to its further development, e.g. [65, 97]. A similar idea was present in [118].

The tool [118] is an integrated environment for designing and verifying SPHINX [1] knowledge bases. It provides a graphical user interface supporting creation of the knowledge base, along with local dynamic verification. The approach is oriented towards application for backward-chaining systems based on a simple attributive language. Further, its verification capabilities are limited mostly to local properties of the so-called decision units.

The main differences between the XTT approach and the one of [118] are that the approach found in KBBUILDER is oriented towards applications for backward-chaining systems, designed by SPHINX/CAKE [86] tools. They are based on simple attributive language and Horn-like clauses. Furthermore, its verification capabilities are limited mostly to local properties of decision units. On the other hand, XTT method is oriented towards forward-chaining systems, and provides a more expressive attributive language, along with a visual knowledge representation and design method. MIRELLA uses high-level PROLOG representation to encode and analyze the rule base. KBBUILDER uses low-level C++ implementation. Moreover, the set and definitions of the formal properties are a bit different with respect to forward-chaining systems in comparison with the ones considered here.

Automated computer tools performing verification of formal properties are not that common as general expert systems development toolkits. An overview of several well-known knowledge-based systems verification and validation suites can be found in [134]. A discussion of rule-based systems verification tools is contained in [118]. Selected important tools are briefly introduced below.

A.1.4 KRUST

The system [26] refines rules considering rule priority and considering the flow of control. The goal of the system is to identify possible faults and anomalies by rule base refinement. KRUST uses training examples to refine the system being analyzed.

A.1.5 IN-DEPTH

It is an incremental verifier that can perform the incremental verification of a knowledge base [85]. It was built to verify knowledge bases designed with MILORD expert systems development environment.

A.1.6 COVER

The system [108] uses multiple advanced verification techniques. Some of the verification algorithms are implemented in PROLOG. However, the verification is possible only after a translation to COVER-specific language.

A.2 Expert Systems Shells

An *expert system shell* provides an inference engine with a user interface, and supports the building of a system knowledge base. Some of the most important shells are presented below [46].

A.2.1 OPS5

It is a classical rule-based language [12, 34]. It has simple inference control algorithms and does not support complex data structures such as graphs or trees. OPS 83 is a successor of OPS5. It is written in C and allows for the integration of applications written in C. OPS 83 supports generalized forward chaining. While it is currently rarely used, it has given foundation to more advanced rule-based languages.

A.2.2 CLIPS

It is a rule-based object-oriented language [41]. CLIPS supports multiple reasoning and conflict resolution strategies. It is one of the most common expert system development tools. CLIPS is an expert system shell, so it does not provide any tools supporting the design of the knowledge base.

A.2.3 Jess

The name stands for *Java Expert System Shell* [35]. It is inspired by CLIPS but implemented in JAVA. Compared to CLIPS it adds several features and offers superior performance. It is easy to integrate with Java-based web-enabled applications.

A.2.4 Sphinx

SPHINX [1] is an integrated environment for expert systems development. It uses backward-chaining inference engine, contains a shell (PCSHELL [87]) and design tools named CAKE [86]. CAKE[2] supports the process of knowledge base design and simple verification.

A.2.5 Oryx/Mandarax

MANDARAX [29] is an open source JAVA class library for deduction rules. It provides an infrastructure for defining, managing and querying rule bases. MANDARAX includes open APIs to interface with relational databases and XML, in particular RuleML. Oryx is a graphical user interface application to design and maintain MANDARAX knowledge bases.

A.2.6 G2

The system [40] is perhaps the most advanced tool for large-scale developments. It is an object-oriented graphical customizable software platform for rapidly building expert manufacturing applications. It allows for building hierarchical models of intelligent systems and for using mixed inference techniques. It provides advanced tools and methods for data acquisition, sharing, and management.

A.2.7 XpertRule

The tool [4] supports developing rule-based systems. It uses a simple visual knowledge builder which maps knowledge modules to decision trees, which constitute main knowledge representation units. It also provides additional features, such as fuzzy reasoning.

A.2.8 ILOG

This integrated environment supports the development and optimization of expert systems [45]. It uses internal knowledge representation language to describe rule-based system. ILOG products contain multiple development tools, including ILOG JRULES, a JAVA and XML-based library.

[2] CAKE is a registered trademark of AITech Artificial Intelligence Laboratory, Katowice, Poland. The name stands for *Computer-Aided Knowledge Engineering*.

A.3 Experimental Systems and New Developments

A.4 IxTeT System

IxTeT is another advanced tool developed for dealing with representation and analysis of time-dependent knowledge. Its main functionality consists of representation and dealing with temporal knowledge representation for monitoring dynamic changes. It can follow prespecified sequences of events in order to ensure that a sequence is properly followed. It can also detect some fault situations as specific predefined situations.

A.5 The Qualitative Engine CA-EN

Ca-En is a universal system for simulation and consistency-based diagnosis of dynamic systems with use of qualitative models. The main application include qualitative simulation; it can also be applied for partial diagnostic inference based on inconsistency detection. A more detailed presentation can be found in [127].

A.6 TIGER: a Real-Time Gas Turbine Monitoring System

This is a large, real-domain application in knowledge-based monitoring, supervision, and diagnosis. The system operates on-line, 24 hours a day, and is applied for continuous monitoring, situation assessment and diagnosis of gas turbines. Its distinctive features include application of the above tools, i.e. Kheops, IxTeT, and Ca-En, systems, i.e. it is a multi-strategy, multi-component system. Details about the Tiger system can be found in the literature quoted above (with respect to its components) and overall presentations are in recent presentation of the state-of-the-art concerning the Tiger methodology and applications can be found in [127] and [88].

A.7 RuleML

RuleML [9] is an XML-based rule markup language devoted to knowledge representation issues. RuleML encompasses a hierarchy of rules, including reactive rules (event-condition-action rules), transformation rules (functional-equational rules), derivation rules (implicational-inference rules), also ones restricted to facts ('premiseless' derivation rules) and queries ('conclusionless' derivation rules), as well as integrity-constraints (consistency-maintenance rules).

The RuleML hierarchy of general rules branches into the two direct categories of reaction rules and transformation rules. On the next level, transformation rules specialize to the subcategory of derivation rules. Then, derivation rules have further subsubcategories, namely facts and queries. Finally, queries specialize to integrity constraints.

In order to represent different rule syntax and semantics RuleML has several *dialects*, or *sublanguages*, including: Datalog and Hornlog. The former is suitable for representing rules equivalent to Horn clauses.

A.8 VisiRule

VisiRule is a graphical tool for designing, developing and delivering business rule and decision support applications. The user can draw a flowchart that represents the decision logic. The main components of this tool are constituted by several predefined blocks, and the crucial components are *Question Boxes* with various outputs defining answer possibilities. The system operates under Windows in the WIN-PROLOG environment of LPA (*Logic Programming Associates*. More information can be found in the WWW page `http://www.lpa.co.uk`.

B

Selected Web Resources

In this appendix a list of some selected resources concerning rule-based systems and tools is provided[1].

B.1 Expert and Rule-Based Systems Resources

Expert System Shells

The site contains a variety of Expert System and Production Systems resources.
http://www-2.cs.cmu.edu/afs/cs/project/ai-repository/ai/areas/expert/0.html

PD OPS5

Public domain implementation of an OPS5 interpreter.
http://www.idiom.com/free-compilers/TOOL/OPS5-1.html

CLIPS

Main page of CLIPS expert system shell.
http://www.ghg.net/clips/CLIPS.html

InfoSapient

INFOSAPIENT is an Open Source for business rule management implemented in JAVA. It employs XML for knowledge and rule representation.
http://info-sapient.sourceforge.net

[1] The list is provided courtesy of Grzegorz Jacek Nalepa, Ph.D. and it is extracted from his Ph.D. Thesis [92].

A. Ligęza: *Logical Foundations for Rule-Based Systems*, Studies in Computational Intelligence (SCI) 11, 289–295 (2006)
www.springerlink.com

JEOPS

JEOPS is a rule-based inference engine extending JAVA with a forward-chaining inference engine.
`http:www.di.ufpe.br/~jeops`

Jess

JESS is a rule engine and scripting environment written entirely in Sun's JAVA language by Ernest Friedman-Hill at Sandia National Laboratories in Livermore, CA. JESS was originally inspired by the CLIPS expert system shell, but has grown into a complete, distinct, dynamic environment of its own.
`http://herzberg.ca.sandia.gov/jess`

Mandarax and Oryx

MANDARAX is an open source JAVA class library for business (deduction) rules. It provides an infrastructure for defining, managing and querying rule bases.
`http://mandarax.sourceforge.net`

ORYX STANDALONE is a graphical user interface application to design and maintain MANDARAX knowledge bases.
`http://www.jbdietrich.de`

OFBiz Rule Engine

OFBIZ is a set of tools for business applications. It contains PROLOG-based inference engine.
`http://www.ofbiz.org`

W4 Project

The W4 project aims at developing Standard Prolog inter-operable tools for supporting distributed, secure, and integrated reasoning activities in the Semantic Web.
`http://centria.di.fct.unl.pt/~cd/projectos/w4`

B.2 RBS-related XML Resources

RuleML

Rule Markup Language main page.
`http://www.ruleml.org`

Essential RuleML

A primer on RuleML.
`http://www.ruleml.org/submission/essentialruleml.html`

LogicML

LogicML is a simple rules markup language for reasoning on the web and interchanging rules. RuleML was the primary reference language in creating LogicML. LogicML includes elements for representing ruleflows. Ruleflow is a distinct feature of commercial rule engine systems, which allows people to author rule-base in the aspect of decision sequence.
`http://machine-knows.etri.re.kr/bossam/docs/logicml.html`

B.3 Selected AI Links

CMU Artificial Intelligence Repository

The AI Repository was established by Mark Kantrowitz in 1993 to collect free software and materials of general interest to AI researchers, educators, students, and practitioners.
`http://www-2.cs.cmu.edu/afs/cs.cmu.edu/project/ai-repository/ai/0.html`

AI Depot

News, knowledge and discussion for the AI enthusiasts.
`http://ai-depot.com/`

AboutAI

An AI portal.
`http://www.aboutai.net`

Artificial Intelligence, History, Philosophy and Practice

A comprehensive collection of links.
`http://www.tau.ac.il/humanities/philos/ai`

Generation5

An AI portal.
`http://www.generation5.org`

KurzweilAI.Net

KurzweilAI.net features the big thoughts of today's big thinkers examining the confluence of accelerating revolutions that are shaping our future world, and the inside story on new technological and social realities from the pioneers actively working in these areas.
`http://www.kurzweilai.net`

AI and VV

Comprehensive AI and verification and validation links.
`http://membres.lycos.fr/hgwet/aistuff.htm`

AI Wiki

This is a collaboratively created and edited area dedicated to all facets of Artificial Intelligence
`http://www.ifi.unizh.ch/ailab/aiwiki`

eBook2U AI

A large collection of references to AI resources, projects and tools.
`http://www.ebook2u.com/web/Computers/Artificial_Intelligence`

PC AI

PCAI Artificial Intelligence: Free eMagazine, White Papers, Demos, Products, Glossary, Links
`http://www.pcai.com`

B.4 Selected Prolog Compilers and Environments

SWI Prolog

A very popular, nice and powerful PROLOG compiler. SWI-PROLOG is a Free Software Prolog compiler, licensed under the Lesser GNU Public License. Together with its graphics toolkit XPCE, its development started in 1987 and has been driven by the needs for real-world applications. Being free, small and standard compliant, SWI-PROLOG has become very popular for education.
`http://www.swi-prolog.org`

GNU Prolog

GNU PROLOG is a free PROLOG compiler with constraint solving over finite domains developed by Daniel Diaz. GNU PROLOG accepts PROLOG+ constraint programs and produces native binaries (like gcc does from a C source). The obtained executable is then stand-alone. The performances of GNU PROLOG are very encouraging (comparable to commercial systems). Beside the native-code compilation, GNU PROLOG offers a classical interactive interpreter (top-level) with a debugger. The PROLOG part conforms to the ISO standard for PROLOG with many extensions very useful in practice. GNU PROLOG also includes an efficient constraint solver over Finite Domains (FD). This opens constraint logic programming to the user combining the power of constraint programming to the declarative nature of of logic programming.
`http://pauillac.inria.fr/~diaz/gnu-prolog/`

XSB

XSB is a Logic Programming and Deductive Database system for Unix and Windows. It is being developed at the Computer Science Department of the Stony Brook University, in collaboration with Katholieke Universiteit Leuven, Universidade Nova de Lisboa, Uppsala Universitet and XSB, Inc. XSB is licensed under GNU Lesser General Public License.
`http://xsb.sourceforge.net`

Amzi! Prolog + Logic Server

Offers embedding PROLOG rule-based components in C/C++, JAVA, DELPHI, VISUAL BASIC, Web Servers (Servlets, JSP, ASP.NET, CGI) and more; developing Unicode and/or ASCII logicbases; using the AMZI! ECLIPSE IDE with source code debugger for local, embedded and remote PROLOG components. Free edition (180 days single PC license) is available.
`http://www.amzi.com`

LPA Prolog

LPA PROLOG is a modern PROLOG compiler and environment operating under Windows. They offer also an expert system shell and a visual editor named VISIRULE.
`http://www.lpa.co.uk`

B.5 Books and Tutorials

Logic, Programming and Prolog

The classic book on logic programming by Ulf Nilsson and Jan Maluszynski, previously published by John Wiley and Sons Ltd.
`http://www.ida.liu.se/~ulfni/lpp`

Adventure in Prolog

The book by Dennis Merritt, published on-line by Amzi! Inc.
`http://www.amzi.com/AdventureInProlog/advtop.htm`

Building Expert Systems in Prolog

The book by Dennis Merritt, published on-line by Amzi! Inc.
`http://www.amzi.com/ExpertSystemsInProlog/xsiptop.htm`

Prolog Programming A First Course

The course by Paul Brna is intended for undergraduate students who have some programming experience and may even have written a few programs in PROLOG.
`http://cblpc0142.leeds.ac.uk/~paul/prologbook`

Prolog programming

An on-line guide to PROLOG by Roman Bartak.
`http://kti.mff.cuni.cz/~bartak/prolog`

Prolog tutorial

A very comprehensive tutorial by J.R.Fisher.
`http://www.csupomona.edu/~jrfisher/www/prolog_tutorial/contents.html`

Quick Prolog

An introductory book about PROLOG.
`http://www.dai.ed.ac.uk/groups/ssp/bookpages/quickprolog/quickprolog.html`

Learn Prolog Now

An on-line PROLOG course.
`http://www.coli.uni-sb.de/~kris/prolog-course`

B.6 Selected Resources

WWW Library

Virtual Library The World Wide Web, Logic Programming resources and links.
`http://vl.fmnet.info/logic-prog`

Prolog Information

PROLOG programming Information.
`http://www.programming-x.com/programming/prolog.html`

Logic Programming

The web page is devoted to the development of the use of logic programming and PROLOG world-wide.
`http://www.logic-programming.org`

Prolog Links

A resource page for PROLOG programmers.
`http://www.codebox.8m.com/prolog.htm`

CMU Prolog Repository

The PROLOG Repository is part of the CMU Artificial Intelligence Repository. The goal of the PROLOG Repository is to collect files and programs of general interest to PROLOG programmers. Information files include the FAQ (Frequently Asked Questions) postings for the comp.lang.prolog newsgroup and copies of the draft standard for PROLOG.
`http://www-2.cs.cmu.edu/afs/cs.cmu.edu/project/ai-repository/ai/lang/prolog/0.html`

AI Logic Programming

`http://www.pcai.com/web/ai_info/logic_programming.html`

B.G. Mirella

At the following address some details on MIRELLA, the system mentioned in this book are available.
`http://mirella.ia.agh.edu.pl`

References

[1] Aitech Katowice. Sphinx 2.3. `http://www.aitech.gliwice.pl/`.

[2] H. R. Andersen. An introduction to binary decision diagrams. Lecture notes for 49285 advanced algorithms E97, Department of Information Technology, Technical University of Denmark, `http://www.it.dtu.dk/~hra`, 1997.

[3] E. P. Andert. Integrated knowledge-based system design and validation for solving problems in uncertain environments. *Int. J. of Man-Machine Studies*, 36:357–373, 1992.

[4] Attar Software. Xpertrule 3.0. `http://www.attar.com/pages/info_xr.htm`.

[5] E. Awad. *Building Expert Systems.* West Publishing Co., 1996.

[6] A. Bendou and M. Ayel. Validation of rule bases containing constraints. *ECAI'96 Workshop on Validation, Verification and Refinement of Knowledge-Based Systems*, pp. 120–125, 1996.

[7] W. Bibel. *Automated Theorem Proving.* Friedr. Vieweg & Sohn, Braunschweig /Wiesbaden, 1982.

[8] W. Bibel and E. Eder. *Methods and calculi for deduction*, a Chapter in [36], pp. 68–182. 1993.

[9] H. Boley, S. Tabet and G. Wagner. Design rationale of RuleML: A markup language for semantic web rules. In *SWWS'01*, Stanford, 2001.

[10] K. A. Bowen. *Prolog and Expert Systems.* Computer Science Series. McGraw-Hill International Editions, 1991.

[11] I. Bratko. *Prolog Programming for Artificial Intelligence.* Addison Wesley, 1986.

[12] L. Brownston, R. Farrell, E. Kant and N. Martin. *Programming Expert Systems in OPS5.* Addison-Wesley, 1985.

[13] Z. Bubnicki: *Uncertain Logics, Variables and Systems.* Springer-Verlag, Berlin, Heidelberg, 2002. LNCIS (Lecture Notes in Control and Infromation Sciences), 276.

[14] B. G. Buchanan *et al. Building Expert Systems*, [43], chapter Constructing an Expert System, pp. 127–167. Addison-Wesley, 1983.

[15] J. Cendrowska. Prism: An algorithm for inducing modular rules. *Int. J. Man -Machine Studies*, 27:349–370, 1987.

[16] C.-L. Chang and R. C.-T. Lee. *Symbolic Logic and Mechanical Theorem Proving.* Academic Press, New York and London, 1973.

[17] A. M. Cheng. *Real-Time Systems. Scheduling, Analysis and Verification.* Wiley-Interscience. John Wiley and Sons, Inc. Publication, Hoboken, New Jersey, 2002.

[18] P. Cichosz. *Systemy uczące się.* Wydawnictwa Naukowo-Techniczne, Warszawa, Poland, 2000 (in Polish).

[19] W. F. Clocksin and C. S. Mellish. *Programming in Prolog* 4th edition. Springer Verlag, September 1994.

[20] F. Coenen. Verification and validation in expert and database systems: The expert systems perspective. *A Keynote presentation in* [61], pp. 16–21, 1998.

[21] F. Coenen. Validation and verification of knowledge based systems: Report on EUROVAV'99. *Knowledge Engineering Review 15:2*, pp. 187–196, 2000.

[22] F. Coenen, B. Eaglestone and M. Ridley. Validation, verification and integrity in knowledge and data base systems: future directions. In [135], pp. 297–311, 1999.

[23] T. Connolly, C. Begg and A. Strachan. *Database Systems, A Practical Approach to Design, Implementation, and Management.* Addison-Wesley, 2nd edition, 1999.

[24] M. A. Covington, D. Nute and A. Vellino. *Prolog Programming in Depth.* Prentice-Hall, 1997.

[25] B. J. Cragun and H. J. Steudel. A decision-table-based processor for checking completeness and consistency in rule-based expert systems. *Int. J. Man-Machine Studies*, 26:633–648, 1987.

[26] S. Craw. Refinement complements verification and validation. *Int. J. Human Computer Studies*, 44(2):245–256, 1996.

[27] R. de Hoog. Methodologies for Building Knowledge-Based Systems: Achievements and Prospectus. A chapter in [51], pp. 1-1–1-14.

[28] J. de Kleer. An assumption based TMS, extending the TMS and problem solving with ATMS. *Artificial Intelligence*, 28(2):127–162, 1986.

[29] J. Dietrich. *The Mandarax Manual.* Massey University, New Zealand, Dec 2003. `http://mandarax.sourceforge.net`.

[30] R. Elmasri and S. B. Navathe. *Fundamentals of Database Systems.* Addison Wesley, 2000.

[31] R. E. Fikes and N. J. Nilsson. Strips: a new approach to the application of theorem proving to problem solving. *Artificial Intelligence*, 2, 1971, 189–208.

[32] P. Flach and A. Kakas (Eds.) *Proceedings of the IJCAI'97 Workshop on Abduction and Induction in AI.* Nagoya, Japan, 1997.

[33] R. Fleurquin and G. Motet. *Wprowadzenie do problematyki jakości oprogramowania. Normy ISO 9000 i programy jakości.* CCATIE, Kraków, 1998 (tytuł oryginału: Introduction à la Qualité Logicielle), tłum. z języka francuskiego A. Ligęza.

[34] C. L. Forgy. *OPS5 User's Manual, Technical Report CMU-CS-81-135.* Carnegie Mellon University, 1981.

[35] E. J. Friedman-Hill. *Jess, The Rule Engine for the Java Platform.* Distributed Computing Systems, Sandia National Laboratories, Livermore, CA, May 2004. `http://herzberg.ca.sandia.gov/jess`.

[36] D. M. Gabbay, C. J. Hogger and J. A. Robinson (Eds). *Handbook of Logic in Artificial Intelligence and Logic Programming*, volume 1 of *Oxford Science Publications.* Clarendod Press, Oxford, 1993.

[37] A. Galton. *Logic for Information Technology.* John Wiley and Sons, Chichester, New York, Brisbane, Toronto, Singapore, 1990.

[38] H. Garcia-Molina, J. D. Ullman and J. Widom. *Database Systems, the Complete Book.* Prentice Hall, 2002.

[39] M. R. Genesereth and N. J. Nilsson. *Logical Foundations of Artificial Intelligence.* Morgan Kaufmann Publishers, Inc., Los Altos, California, 1987.

[40] Gensyn Corporation. G2: Real-time expert systems. `http://www.gensym.com/manufacturing/g2_overview.shtml`, 2004.

[41] J. C. Giarratano. *CLIPS User's Guide, version 6.20*, March 2002. `http://www.ghg.net/clips`.

[42] J.-P. Gouyon. Kheops users's guide. *Report of Laboratoire d'Automatique et d'Analyse des Systemes* (92503), 1994.

[43] F. Hayes-Roth, D. A. Waterman and D. B. Lenat (Eds). *Building Expert Systems.* Addison-Wesley, Reading, Massachusetts, 1983.

[44] A. A. Hopgood. *Intelligent Systems for Engineers and Scientists.* CRC Press, Boca Raton London New York Washington, D.C., 2nd edition, 2001.

[45] Ilog Inc. Ilog rules. `http://www.ilog.co.uk/products/rules`, 2004.

[46] P. Jackson. *Introduction to Expert Systems.* Addison–Wesley, 3rd edition, 1999.

[47] R. Kowalski. Predicate logic as a programming language. In *Proceedings of IFIP-74*, pp. 569–74, 1974.

[48] T. Laffey and *et al.* Real-time knowledge-based systems. *AI Magazine*, Spring: 27–45, 1988.

[49] N. Lamb and A. Preece. Verification of multi-agent knowledge-based systems. *ECAI'96 Workshop on Validation, Verification and Refinement of Knowledge-Based Systems*, pp. 114–119, 1996.

[50] P. E. Lehner. An overview of intelligent systems technology. In G. W. H. Stephen J. Andriole (Eds), *Applied Artificial Intelligence A Sourcebook*, Chap. 1. McGraw-Hill, Inc., 1992.

[51] J. Liebowitz. *The Handbook of Applied Expert Systems.* CRC Press, 1998.

[52] J. Liebowitz. *Knowledge Management, Learning from Knowledge Engineering.* CRC Press, 2001.

[53] A. Ligęza. Logical foundations for knowledge-based control systems — knowledge representation, reasoning and theoretical properties. *Scientific Bulletins of AGH*, Automatics, 63(1529):144 pp., Kraków, 1993.

[54] A. Ligęza. A note on backward dual resolution and its application to proving completeness of rule-based systems. *Proceedings of the 13th Int. Joint Conference on Artificial Intelligence (IJCAI), Chambery, France*, 1:132–137, 1993.

[55] A. Ligęza. Backward dual resolution. direct proving of generalization. *Information Modeling and Knowledge Bases V: Principles and formal techniques*, pp. 336–349, 1994.

[56] A. Ligęza. Logical foundations for knowledge-based control systems, part I: Language and reasoning. *Archives of Control Sciences*, 3(XXXIX)(3-4):289–315, 1994.

[57] A. Ligęza. Logical foundations for knowledge-based control systems. part II: Representation of states, transformations, and analysis of theoretical properties. *Archives of Control Sciences*, 4(XL)(1-2):129–166, 1994.

[58] A. Ligęza. Towards design of complete rule-based control systems. In R. K. J. Kocijan (Ed), *IFAC/IMACS International Workshop on Artificial Intelligence in Real-Time Control*, pp. 189–194. IFAC, Bled, Slovenia, 1995.

[59] A. Ligęza. Logical support for design of rule-based systems. Reliability and quality issues. In M. Rousset (Ed.), *ECAI-96 Workshop on Validation, Verification and Refinment of Knowledge-based Systems*, volume W2, pp. 28–34. ECAI'96, Budapest, 1996.

[60] A. Ligęza. Logical analysis of completeness of rule-based systems with dual resolution. *EUROVAV'97 — 4th European Symposium on the Validation and Verification of Knowledge Based Systems*, pp. 19–29, 1997.

[61] A. Ligęza. Towards logical analysis of tabular rule-based systems. *Proceedings of the Ninth European International Workshop on Database and Expert Systems Applications*, Vienna, Austria, pp. 30–35, 1998.

[62] A. Ligęza. Intelligent data and knowledge analysis and verification; towards a taxonomy of specific problems. *Proceedings of the 5-th European Symposium on Verification and Validation of Knowledge Based Systems and Components: Validation and Verification of Knowledge Based Systems: Theory, Tools and Practice, EUROVAV'99*, 1999.

[63] A. Ligęza. Intelligent data and knowledge analysis and verification; towards a taxonomy of specific problems. *Validation and Verification of Knowledge Based Systems: Theory, Tools and Practice*, pp. 313–325, 1999.

[64] A. Ligęza. Elements of algebraic data analysis for verification of qualitative properties. *Proceedings of the Conference* Knowledge Acquisition from Databases, pp. 18–28, 2000.

[65] A. Ligęza. Toward logical analysis of tabular rule-based systems. *International Journal of Intelligent Systems*, pp. 333–360. 2001.

[66] A. Ligęza *et al.* Supervision systems. `http://eia.udg.es/~iitap/monografia/index-eng.html`.

[67] A. Ligęza and P. Fuster Parra. Towards logical analysis of rule-based systems. *Proceedings of the 13th European Meeting on Cybernetics and Systems Research*, 2:1211–1216, 1996.

[68] A. Ligęza, G. J. Nalepa and I. Wojnicki. Analysis of selected problems of design and implementation of real-time rule-based systems on the base of Kheops system (in Polish). In T. Szmuc and R. Klimek (Eds), *Real Time Systems 2000*, pp. 53–64. Institute of Automatics AGH, Cracow, 2000.

[69] A. Ligęza, I. Wojnicki and G. J. Nalepa. Tab-trees: a case tool for design of extended tabular systems. In H. M. et al. (Eds), *Database and Expert Systems Applications*, volume 2113 of *Lecture Notes in Computer Sciences*, pp. 422–431. Springer-Verlag, Berlin, 2001.

[70] A. Ligęza, I. Wojnicki and G. J. Nalepa. Design and implementation support of rule-based systems (in polish). In K. Zieliński (Ed), *II Polish National Conference on Software Engineering*, pp. 229–236. AGH, Zakopane/Cracow, 2000.

[71] A. Ligęza. An introduction to knowledge-based process monitoring, supervision and diagnosis. Basic ideas, problems and theoretical foundations. Working notes. 1997.

[72] A. Ligęza. Granular algebra: Towards a calculus of semi-partitions for analysis, manipulation and verification of tabular systems. In R. Trappl (Ed), *Cybernetic and Systems*, volume 2 of *Proceedings of the Sixteenth European*

Meeting on Cybernetics and Systems Research, pp. 806–811, Vienna, 2002. University of Vienna and Austrian Society of Cybernetic Studies.

[73] A. Ligęza. Granular sets and granular relations. An algebraic approach to knowledge representation and reasoning. In T. Burczyński, W. Cholewa, and W. Moczulski (Eds), *Methods of Artificial Intelligence*, Proceedings of the Symposium on Methods of Artificial Intelligence *AI-METH 2002*, pp. 47–54, Gliwice, Poland, Silesian University of Technology 2002.

[74] A. Ligęza. Granular sets and granular relations. Towards a higher abstraction level in knowledge representation. In M. A. Kłopotek, S. T. Wierzchoń, and M. Michalewicz (Eds), *Intelligent Information Systems 2002*, Advances in Soft Computing, pp. 331–340, Heidelberg and New York, 2002. Physica-Verlag. A Springer Verlag Company.

[75] A. Ligęza. Dual resolution for logical reduction of granular tables. In M. A. Kłopotek, S. T. Wierzchoń, and K. Trojanowski (Eds), *Intelligent Information Processing and Web Mining. Proceedings of the International IIS: IIPWM'03 Conference held in Zakopane, Poland, June 2–5, 2003*, Advances in Soft Computing, pp. 363–372, Berlin, Heidelberg, 2003. Springer-Verlag.

[76] A. Ligęza. Generalized decision tables as a tool for knowledge management. Selected issues of knowledge representation, analysis and processing. In M. Nycz and M. L. Owoc (Eds), *Pozyskiwanie Wiedzy i Zarządzanie Wiedzą*, number 975 in Prace Naukowe Akademii Ekonomicznej im. Oskara Langego we Wrocławiu, pp. 279–290, Wrocław, 2003. Akademia Ekonomiczna im. Oskara Langego we Wrocławiu.

[77] A. Ligęza. Granular sets and granular relations. Applications to knowledge representation and tabular systems analysis. *Automatyka*, 7(1–2):141–146, 2003.

[78] A. Ligęza. Granular sets and granular relations for algebraic knowledge management. In C. A. Dagli, A. L. Buczak, J. Ghosh, M. J. Embrechts, and O. Ersoy (Eds), *Smart Engineering System Design: Neural Networks, Fuzzy Logic, Evolutionary Programming, Complex Systems, and Artificial Life. Proceedings of the Artificial Neural Networks in Engineering Conference (ANNIE 2003), St. Louis, Missouri*, volume 13 of *Intelligent Engineering Systems through Artificial Neural Networks*, pp. 169–174, New York, 2003. ASME Press.

[79] A. Ligęza. Logical reduction of tabular systems. In M. Nycz and M. L. Owoc (Eds), *Pozyskiwanie wiedzy i zarządzanie wiedzą*, no. 1011 in Prace Naukowe Akademii Ekonomicznej im. Oskara Langego we Wrocławiu, pp. 188–199. Wydawnictwo Akademii Ekonomicznej im. Oskara Langego we Wrocławiu, Wrocław, 2004.

[80] A. Ligęza and M. Szpyrka. Reduction of tabular systems. In L. Rutkowski, J. Siekmann, R. Tadeusiewicz and L. A. Zadeh (Eds), *Artificial Intelligence and Soft Computing. Proceedings of the 7-th International Conference, Zakopane, Poland, June 2004*, volume LNAI 3070 of *Lecture Notes in Artificial Intelligence*, pp. 903–908, Berlin, Heidelberg, New York, Springer 2004.

[81] A. D. Lunardhi and K. M. Passino. Verification of qualitative properties of rule-based expert systems. *Applied Artificial Intelligence*, 9:587–621, 1995.

[82] D. Maier and D. S. Warren. *Computing with logic. Logic Programming with Prolog.* The Benjamin/Cummings, 1988.

[83] W. Marek. Basic properties of knowledge base systems. *The Knowledge Frontier*, pp. 137–160, 1987.

[84] D. Merritt. *Building Expert Systems in Prolog*. Springer-Verlag, 1989. on live version: `http://www.amzi.com/ExpertSystemsInProlog`.

[85] P. Meseguer and A. Verdaguer. Verification of multi-level rule-based expert systems: Theory and practice. *International Journal of Expert Systems*, 6(2):163–192, 1993.

[86] K. Michalik. *CAKE 4.0, Komputerowy System Wspomagania Inżynierii Wiedzy*. AITech Artificial Intelligence Laboratory, Katowice, Poland, 2003.

[87] K. Michalik. *PC-Shell 4.0, Szkieletowy System Ekspertowy*. AITech Artificial Intelligence Laboratory, Katowice, Poland, 2003.

[88] R. Milne and C. Nicol. Tiger: Continuous diagnosis of gas turbines. In W. Horn (Ed), *ECAI'2000. 14th European Conference on Artificial Intelligence*, pp. 711–715, Amsterdam, Berlin, Oxford, Tokyo, Washington DC, 2000. European Coordination Committee on Artificial Intelligence (ECCAI), IOS Press.

[89] G. J. Nalepa. Graphical user interface to kheops rule-based expert system. Master's Thesis, AGH, Kraków, 1999.

[90] G. J. Nalepa. Przegląd wybranych platform programowych pod katem przydatności dla weryfikacji własności baz danych i baz wiedzy. Raport 92, Katedra Automatyki AGH, Kraków, 1999.

[91] G. J. Nalepa. Przegląd własności wybranych współczesnych implementacji języka Prolog dla analizy i weryfikacji baz danych i baz wiedzy. In Z. Bubnicki and A. Grzech (Eds), *Inżynieria Wiedzy i Systemy Ekspertowe*, volume 2, pp. 27–34, Oficyna Wydawnicza Politechniki Wrocławskiej, Wrocław, 2000.

[92] G. J. Nalepa. *Meta-Level Approach to Integrated Design and Implementation of Rule-Based Systems*. Ph.D. Thesis, Kraków, Poland, AGH-UST, 2004.

[93] G. J. Nalepa and A. Ligęza. Graphical case tools for integrated design and verification of rule-based systems. In T. Burczyński, W. Cholewa, and W. Moczulski (Eds), *Methods of Artificial Intelligence*, Proceedings of the Symposium on Methods of Artificial Intelligence *AI-METH 2002*, pp. 307–3013, Gliwice, Poland, 2002. Silesian University of Technology.

[94] G. J. Nalepa and A. Ligęza. Designing reliable rule-based systems with integrated case tools. *Automatyka*, 7(1–2):179–184, 2003.

[95] G. J. Nalepa and A. Ligęza. Integrated design environment for formal verification of rule-based systems. In Z. Bubnicki and A. Grzech (Eds), *Inżynieria Wiedzy i Systemy Ekspertowe*, volume II, pp. 30–37, Wrocław, Oficyna Wydawnicza Politechniki Wrosławskiej 2003.

[96] G. J. Nalepa and A. Ligęza. Meta-level approach to knowledge management in rule-based systems. In M. Nycz and M. L. Owoc (Eds), *Pozyskiwanie Wiedzy i Zarządzanie Wiedzą*, number 975 in Prace Naukowe Akademii Ekonomicznej im. Oskara Langego we Wrocławiu, pp. 332–339, Wrocław, Akademia Ekonomiczna im. Oskara Langego we Wrocławiu 2003.

[97] G. J. Nalepa and A. Ligęza. Designing reliable web security systems using rule-based approach. In *Advanced in Web Intelligence*, AWIC 2003, pp. 124–133, Madrid, Spain, LNAI2663.

[98] G. J. Nalepa and A. Ligęza. A graphical tabular model for rule-based logic programming and verification. In Z. Bubnicki and A. Grzech (Eds), *Proceedings of the 15-th International Conference on Systems Science*, volume III, pp. 23–28, Wrocław, Oficyna Wydawnicza Politechniki Wrocławskiej 2004.

[99] G. J. Nalepa and A. Ligęza. Markup-languages based approach to knowledge management and representation. In M. Nycz and M. L. Owoc (Eds), *Pozyskiwanie wiedzy i zarządzanie wiedzą*, number 1011 in Prace Naukowe Akademii Ekonomicznej im. Oskara Langego we Wrocławiu, pp. 234–240. Wrocław, Wydawnictwo Akademii Ekonomicznej im. Oskara Langego we Wrocławiu 2004.

[100] G. J. Nalepa and A. Ligęza. Conceptual modelling and automated implementation of rule-based systems. Proceedings of the Polish National Conference of Software Engineering KKIO'2005, IOS Press, Kraków 2005.

[101] D. L. Nazareth. Issues in the verification of knowledge in rule-based systems. *Int. J. Man-Machine Studies*, 30:255–271, 1989.

[102] M. Negnevitsky. *Artificial Intelligence. A Guide to Intelligent Systems.* Harlow, England; London; New York, Addison-Wesley 2002.

[103] T. A. Nguyen and *et al.* Checking an expert systems knowledge base for consistency and completeness. *Proceedings of the 9-th IJCAI*, pp. 375–378, 1985.

[104] N. Nilsson. *Principles of Artificial Intelligence.* Tioga Publishing Co., Palo Alto, California, 1980.

[105] U. Nilsson and J. Małuszyński. *Logic, Programming and Prolog.* John Wiley & Sons, 1990.

[106] Z. Pawlak. *Rough Sets. Theoretical Aspects of Reasoning about Data.* Dordrecht/Boston/London, Kluwer Academic Publishers 1991.

[107] W. A. Perkins and *et al.* Knowledge base verification. *Topics in Expert System Design*, pp. 353–376, 1989.

[108] A. Preece. Methods for verifying expert system knowledge base. Available from apreececsd.abdn.ac.uk.

[109] A. D. Preece. Methods for verifying expert system knowledge bases. *Technical Report*, 37 pages, 1991.

[110] A. D. Preece. A new approach to detecting missing knowledge in expert system rule bases. *Int. J. of Man-Machine Studies*, 38:161–181, 1993.

[111] A. D. Preece *et al.* Verifying rule-based systems. *Technical Report*, 25 pages, 1991.

[112] A. D. Preece and R. Shinghal. Foundation and application of knowledge base verification. *Technical Report*, 26 pages, 1994.

[113] N. Rakoto-Ravalontsalama and J. Aguilar Martin (Eds.). *Supervision de Processus à l'Aide du Système Expert G2.* Hermes, Paris, 1995.

[114] J. A. Robinson. A machine-oriented logic based on the resolution principle. *J. Association for Computing Machinery*, pp. 23–41, 12, 1965.

[115] K. A. Ross and C. R. B. Wright. *Discrete Mathematics*, 3rd Edition. Prentice Hall Inc., 1992.

[116] M. T. Saborido. An introduction to expert system development. In R. A. V. L. Boullart, A. Krijgsman (Eds), *Application of Artificial Intelligence in Process Control.* Pergamon Press, 1992.

[117] A. Shinghal. *Formal Concepts in Artficial Intelligence.* 1st edition, Chapman & Hall, London, 1992.

[118] R. Simiński. *Dynamiczna weryfikacja poprawności baz wiedzy w procesie ich weryfikacji.* Ph.D. Thesis, Instytut Podstaw Informatyki PAN, Warszawa, 2002.

[119] A. Sinton. A safety analysis of the airbus a320 braking system design. Master's thesis, Department of Computing and Mathematics, University of Stirling, 1994.

[120] H. T. H. Solomon L. Pollack and W. J. Harrison. *Decision Tables: Theory and Practice.* Wiley-Interscience, a Division of John Wiley and Sons, Inc., New York, London, 1971.

[121] L. Sterling and E. Shapiro. *The Art of Prolog.* Advanced Programming Techniques (Logic Programming). MIT Press, March 10 1994.

[122] M. Stock. *AI in Process Control.* Intertext Publications/Multiscience Press, Inc., New York, 1989.

[123] M. Suwa, C. A. Scott, and E. H. Shortliffe. Completeness and consistency in rule-based expert system. *Rule-Based Expert Systems*, pp. 159–170, 1985.

[124] J. Tepandi. Verification, testing, and validation of rule-based expert systems. *Proceedings of the 11-th IFAC World Congress*, pp. 162–167, 1990.

[125] I. S. Torsun. *Foundations of Intelligent Knowledge-Based Systems.* Academic Press, London, San Diego, New York, Boston, Sydney, Tokyo, Toronto, 1995.

[126] L. Travé-Mesuyèz, R. Milne, C. Nicol and J. Quevedo. Tiger: Knowledge based gas turbine condition monitoring. In C. C. A. Macintosh, editor, *Applications and Innovations in Expert Systems III*, volume III, pp. 23–43. SGES Publications, Cambridge, Oxford, 1995.

[127] L. Travé-Mesuyèz. Gas-turbine condition monitoring using qualitative model-based diagnosis. *IEEE Expert Intelligent Systems & Their Applications*, May/June:22–31, 1997.

[128] S. G. Tzafestas. System fault diagnosis using the knowledge-based methodology. In P. F. R. Patton and R. Clark (Eds), *Fault Diagnosis in Dynamic Systems. Theory and Applications*, pp. 509–572. New York, Prentice Hall International Ltd. 1989.

[129] S. Tzafestas, S. Ata-Doss, and G. Papakonstantinou. Expert system methodology in process supervision and control. In S. Tzafestas (Ed), *Knowledge-Base System Diagnosis, Supervision and Control*, pp. 181–215. New York, London, Plenum Press 1989.

[130] S. Tzafestas, S. Ata-Doss, and G. Papakonstantinou. *Knowledge-Base System Diagnosis, Supervision and Control*, chapter Expert system methodology in process supervision and control, pp. 181–215. New York, London, Plenum Press 1989.

[131] J. D. Ullman. *Principles of Database and Knowledge Base Systems*, volume 1. Computer Science Press, 1988.

[132] F. van Harmelen. Applying rule-based anomalies to kads inference structures. *ECAI'96 Workshop on Validation, Verification and Refinement of Knowledge-Based Systems*, pp. 41–46, 1996.

[133] J. Vanthienen. *PROcedural LOgic Analyzer 5.1*, September 2000.

[134] A. Vermesan. Foundation and Application of Expert System Verification and Validation. CRC Press, 1998. A chapter in [51], pp. 5-1–5-32.

[135] A. Vermesan and F. Coenen (Eds.). *Validation and Verification of Knowledge Based Systems. Theory, Tools and Practice.* Kluwer Academic Publisher, Boston, 1999.

[136] A. Vermesan *et al.* Verification and validation in support for software certification methods. Kluwer Academic Publisher, 1999. A chapter in [135], pp. 277–295.

[137] S. Weiss and C. Kulikowski. *A Practical Guide to Designing Expert Systems.* London, Chapman and Hall Ltd., 1984.

[138] J. Wielemaker. *SWI-Prolog Reference Manual.* Dept. of Social Science Informatics, University of Amsterdam, 2002.

[139] B. J. Wielinga, A. T. Schreiber, and J. A. Breuker. KADS: A modelling approach to knowledge engineering. *Knowledge Acquisition*, 4(1):5–53, 1992. Special issue: The KADS approach to knowledge engineering.

[140] B. J. Wielinga, A. T. Schreiber, and R. de Hoog. *Knowledge and Decisions in Health Telematics*, chapter Modelling perspectives in medical KBS construction. IOS Press, 1994.

[141] I. Wojnicki. Design and implementation of a graphical user interface for computer aided logical design of kheops knowledge based system. Master's Thesis, Academy of Mining and Metallurgy, 2000.

[142] M. L. Wright *et al.* An expert system for real-time control. *IEEE Software*, pp. 16–24, March 1986.

[143] S. Zbroja and A. Ligęza. Case-based reasoning within tabular systems. extended structural data representation and partial matching. *Proceedings of the Conference* Flexible Query Answering Systems, pp. 199–201, 2000.

[144] Z. Zwinogrodzki. *Automatyczne dowodzenie twierdzeń.* Kraków, Wyd. AGH 1976 (in Polish).

Index

MIX
Papier aus verantwortungsvollen Quellen
Paper from responsible sources
FSC® C105338

If you have any concerns about our products,
you can contact us on
ProductSafety@springernature.com

In case Publisher is established outside the EU,
the EU authorized representative is:
Springer Nature Customer Service Center GmbH
Europaplatz 3, 69115 Heidelberg, Germany

Printed by Libri Plureos GmbH
in Hamburg, Germany